CW00497928

ISBN 978-1-332-32136-0
PIBN 10313724

LABORATORY EXERCISES

IN

ELEMENTARY PHYSICS

BY

CHARLES R. ALLEN, S.B.

Instructor in the New Bedford, Mass., High School

NEW YORK
HENRY HOLT AND COMPANY
1892

Robert Drummond,
Electrotyper and Printer,
New York.

PREFACE.

MOST of the experiments in this collection exact of the pupil measurements of some sort, that is, are quantitive. A few, included for the training in accurate work they afford or for their suggestiveness, demand the investigations of the conditions under which certain phenomena develop. They require, however, the use of some physical instrument. Those purely illustrative are given because they demand more careful observation than the pupil can give to a lecture experiment. In the course will be found, I think, illustrations of many of the more common methods of physical research. The exercises are planned for young pupils with no previous training in physics, employ no unduly expensive apparatus, and require no more than forty-five minutes each in the laboratory. The subjects selected are among those bearing on the commoner applications of physical science. Pains have been taken to so frame the instructions that the pupil can prepare himself beforehand to make the most of his laboratory time with the least help from his instructor. This I regard of prime importance. Unless the instructor assures himself before each exercise that the pupils understand what they are to do and how to do it, they will pretty surely exceed the time-limit, and may even make wreck of the whole exercise. Five unprepared pupils require more attention than fifteen who have thoroughly mastered the preliminary work. When entirely new or especially complicated apparatus is to be used, I find it advisable to place a "dummy" set before the class, and spend part or the whole of one period in requiring individual pupils to go through the motions with it, to answer questions on the general method and the special manipulations, and, in case of a complicated calculation, to work out results from imaginary data.

The arrangement and even the phrasing of the material is the outgrowth of much searching for a general form which would be most effective in stimulating clear and independent thinking. Each exercise is introduced by some preliminary explanation and a distinct statement of its object. This puts before the pupil the precise thing he is after, and the general course of his investigation. The manipulation is then described with what some may deem unnecessary minuteness, but I find this minuteness part of the secret of speed and success. I have added questions where I have found them convenient in guiding the pupils' thought, but in no instance, I believe, do they contain their own answer. In a few cases I have given alternative exercises on the same topic, for the purpose of suiting the varying experimental aptitude of pupils.

In the order of subjects, the exercises form a somewhat roughly graded course, from magnetic phenomena, where work is simplest and most stimulative to attention, through experiments involving the measurement of a single value by means of some single instrument, to the more complicated quantitive determinations of Dynamics. This order places the most difficult part of Physics last, where the pupil can bring to his aid, in grasping abstract ideas and performing intricate experiments, the training acquired in the previous parts of the course. But since the instructions in one subject do not assume a previous knowledge of any other, there is nothing to prevent the subjects being taken up in any order desired, though of course practice in mensuration should precede the quantitive exercises.

The book is made up mainly of the author's instructions to his own pupils for their laboratory work. The course of which this forms a part includes also the performance of all necessary descriptive experiments before the class, and the use of a text-book. Before the pupil goes to work in the laboratory at all, he should be given a general idea of how knowledge is acquired experimentally, and the steps involved in carrying on an experimental investigation. The instruction on these points should be illustrated by some simple typical experiments by the teacher. Afterwards the relative order of text-book and laboratory work will naturally depend upon the nature of the laboratory exercises. In exercises involving the study of conditions, as in those on Magnetism, and some of those on Electricity and Heat, the author prefers that the laboratory work precede the text-book work; but in exercises involving definite measurement, as in Specific Gravity and Specific Heat, this order may well be reversed. Certainly, in the determination of a physical law by measurements of two values, as in the exercises on Elasticity or on the Pendulum, the laboratory work should come first.

In order to reduce the volume to a convenient size for the laboratory table, all matter meant chiefly for the instructor is appended to the Teachers' Edition and omitted from the Pupils' Edition, the two being identical in other respects. Appendix A offers general suggestions for arranging the pupil's work and economizing his and the instructor's time, together with directions for applying the ordinary 110-volt Edison electric current to laboratory work, and for the care of mercury. Appendix B gives complete lists of all the apparatus needed for each exercise, hints as to substitutes and duplicates, and itemized estimates of cost with references to dealers' catalogues. Appendix C contains full instructions for making the more important pieces of apparatus. Appendix D furnishes topical references to Avery's "First Principles of Natural Philosophy" and "Elements of Natural Philosophy," to Gage's "Elements of Physics" and "Introduction to Physical Science," and to Hall and Bergen's "Text-Book of Physics," supplemented by hints on conducting the various exercises, the educational purpose each is meant to serve, the degree of accuracy to be expected, etc.

While, so far as I know, none of the experiments will be found elsewhere in exactly their form here, many have been modified from other manuals. No attempt is made to credit each exercise to the source of the original idea. The chief books laid under contribution are Worthington's "Laboratory's Practice," Stewart and Gee's "Physics," Pickering's "Physical Manipulations," the Harvard College Course of Experiments, and Maxwell's "Matter and Motion." The metric system has been employed because it is the language of quantity in physical laboratories, scientific text-books and journals, and the higher scientific manufacturing processes the world over. Pupils learn with such ease to use so much of it as this book requires, that it forms no bar to their progress.

C. R. A.

NEW BEDFORD, MASS., February 1, 1892.

CONTENTS.

MAGNETISM.

CURRENT ELECTRICITY.

MENSURATION.

DENSITY AND SPECIFIC GRAVITY.

HEAT.

PAGE

DYNAMICS.

LIGHT.

SOUND.

APPENDICES.

LABORATORY PHYSICS.

MAGNETISM.

EXERCISE 1.

GENERAL STUDY OF A MAGNET.

EXPERIMENT 1.

Apparatus.—Bar magnet, a piece of steel (knife-blade or knitting-needle), some carpet-tacks; pieces of paper, glass, wood, etc.; copper tacks; a piece of window-glass two or three inches square; iron-filings; larger tack or nail; bottle which may contain the iron-filings or block of wood.

OBJECT.—To compare the results of bringing first a piece of steel and then the magnet near a piece of iron or another piece of steel.

MANIPULATION.—Bring one end of the magnet near a few tacks scattered on a piece of paper. Observe carefully what happens. Repeat two or three times until you are sure that you have noticed all that occurs. Repeat with an ordinary piece of steel. Compare results. Try other bodies in place of the iron. Do you get the same result?

Definitions.—If whenever a body is brought near a piece of iron or steel we obtain the results observed above, that body is called a *magnet.* This name is given it not because it is made of any particular substance or made in any particular shape, but only because when placed under certain conditions, for instance those in Experiment 1, certain things happen that do not happen when other substances are placed under the same conditions. Anything that happens among physical things (things that have weight or take up room) is called a *physical phenomenon*; for example, the behavior of the tack when the magnet was brought near it would be a phenomenon. Placing a body under certain conditions and observing the resultant phenomena is called an *experiment*. A magnet in the form of a straight bar is usually called a *bar magnet*.

EXPERIMENT 2.

OBJECT.—To see (*a*) if contact is necessary to get the results of Experiment 1, and (*b*) the effect of interposing various bodies between the magnet and the iron.

MANIPULATION.—Stand a tack on its head and bring the magnet slowly up to it. Note particularly whether or not the tack moves before the magnet touches it. Repeat the experiment, holding successively a piece of paper, a piece of glass, and a thin piece of wood between the end of the magnet and the tack.

EXPERIMENT 3.

OBJECT.—To see if the magnetism is of the same strength all along the bar; and if not, how it varies at different points.

MANIPULATION.—Lay the magnet on a sheet of paper and dust iron-filings over it all along the bar. Now raise the bar. By the number of filings that adhere to the various parts of the magnet you can form some idea of the distribution of the magnetism.

EXPERIMENT 4.

OBJECT.—To see if the distance between the magnet and the iron produces any effects.

MANIPULATION.—Stand a large tack on its head, *slowly* bring one end of the magnet up to it, and observe the distance between the end of the tack and the end of the magnet when the tack begins to move. Repeat with a small tack.

EXPERIMENT 5.

OBJECT.—To see what happens when iron is brought between the magnet and another piece of iron.

MANIPULATION.—Lay the magnet so that one end projects over the edge of the table. Attach a tack to it and then bring a second tack up to the first. Try other substances in place of the second tack.

EXPERIMENT 6.

OBJECT.—To measure the magnetic pulls at different points of the bar.

MANIPULATION.—Rest the centre of the magnet on top of a bottle or block of wood, as in Fig. 1, and suspend a tack from one end, then carefully attach a second tack to the first, and so proceed until you have as long a chain of tacks as the magnet will hold. Count and record the number of tacks. Remove the chain and repeat from a point about one-half inch nearer the centre of the magnet. Continue these measurements at points about one-half inch apart through the length of the magnet. Make a table of your results as follows:

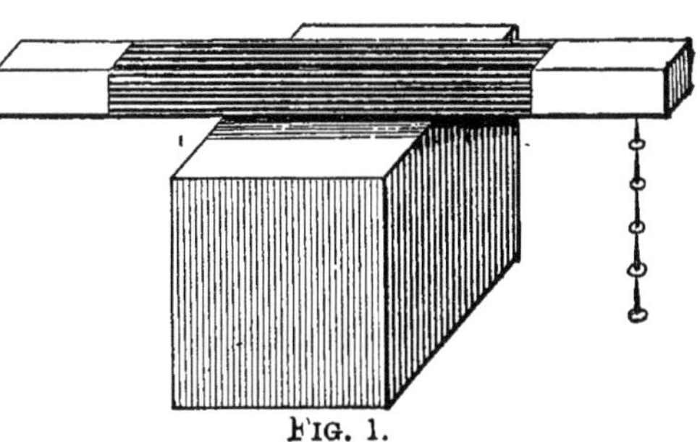

FIG. 1.

Distance from right-hand end of magnet.	Number of tacks.

Holding the magnet vertically, measure in the same way its power at the ends.

Definitions.—The points in the magnet where the magnetism is the strongest are called the *Poles.*

When one body moves towards another, as the tack moved towards the magnet in Exercise 1, it is said to be *attracted;* thus, instead of saying that the tack moved towards the magnet, we would say that the tack was "attracted" by the magnet. Under the same circumstances, if the body moved away, it would be said that it was "*repelled.*"

A *qualitative* experiment is one in which whatever happens is simply observed; a *quantitative* experiment is one in which measurements are made. For example, Experiment 6 in Exercise 1 was quantitative, while all the other experiments were qualitative.

When a body acts as if it were pushed or pulled, as for example the tack in Experiment 1, it is said to be acted on by a *force.*

EXERCISE 2.

THE ACTION OF THE ATTRACTED BODY ON THE MAGNET.

Preliminary.—How the iron behaves toward a magnet was shown in Exercise 1; now it is desired to find out whether the magnet is also affected. When the magnet was brought up to the tack, the tack moved; but the magnet, if it tended to move, could not do so because it was held firmly. In studying the action of the attracted body on the magnet, the conditions of Ex. 1 would naturally be reversed. The magnet would be placed on the table and the tack held near it. This could be done if the magnet were very small, but the magnets ordinarily used are so heavy that, if the experiment were tried in that way, the attraction would have to be very strong in order to move them. If, however, the magnet be suspended, it will not rub against anything if it tends to move, and even a very slight pull will cause it to swing; therefore in the following Exercise the magnet is suspended and a body brought up to it.

EXPERIMENT 1.

Apparatus.—Bar magnet; a new nail or tack; stirrup and thread for suspending the magnet.*

OBJECT.—To see if the attracted body also attracts the magnet.

MANIPULATION.—Make a small stirrup of wire (copper is the best), as in Fig. 2, and by means of it suspend the magnet so that it swings freely and hangs horizontally. When it has come to rest, bring a large nail near one end, but not touching it. Observe carefully what happens, and record.

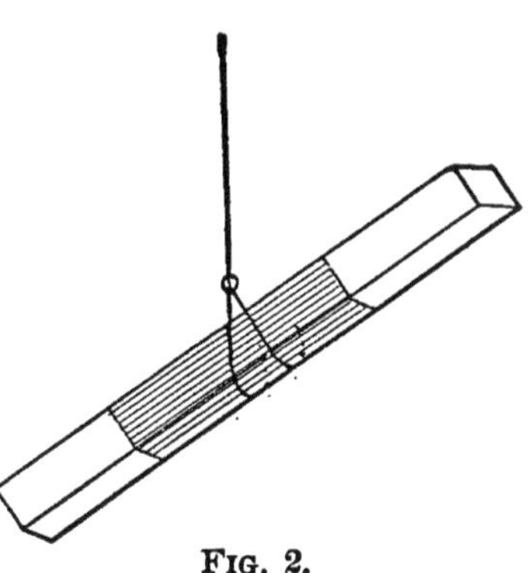

FIG. 2.

Now repeat with the other end of the magnet. Notice particularly whether the results are the same for both ends.

EXPERIMENT 2.

OBJECT.—To see what happens when a magnet is free to move in a horizontal plane.

MANIPULATION.—Set the suspended magnet to swinging gently. Note if it tends to come to rest in any particular direction, and if so in what direction. If there is any doubt, after the magnet has come to rest displace it slightly and see if it shows any tendency to return to that position. Does any particular end of the magnet tend to point in any particular direction?

---

* If a compass is available, it may be conveniently substituted for the suspended magnet.

Definitions.—If something happens to one substance when another substance is brought near it, the second is said to *act* on the first. If the action is mutual, either body may be said to act and the other would then be said to *react*. Does a magnet act on a piece of iron? Is there any reaction? If so, where?

The pole of a magnet that tends to point north is called the North-seeking or *North pole*. In the same way the other is called the *South pole*.

When new properties are developed in one body by bringing another body near it, the new properties of the first body are said to be *induced* by the second.

Any contrivance for getting desired conditions is called a piece of apparatus; for instance, a compass is a contrivance or piece of apparatus for arranging a magnet so that it can swing freely.

When the same experiment has been tried a great many times, with the same results wherever or whenever it was tried, such results are said to be a *law*. For example, whenever a magnet has been brought near a piece of iron that could move, the iron has moved towards the magnet; and whenever the magnet has been placed under such contions that it could move, it has moved toward the iron. Whenever this has not happened, it has always been found that some other conditions had been introduced into the experiment, and when these were removed the usual results have taken place; hence it is said to be a law that a magnet attracts a piece of iron and the piece of iron attracts the magnet.

EXERCISE 3.

MUTUAL ACTION OF TWO MAGNETS.

Preliminary.—So far, magnets have been studied; but in the following Exercise the subjects are not magnets, but magnetic poles, and it is desired to study the action of one magnetic pole on another. In order to do this, one pole must be free to move and the other must be brought near it. As there are two kinds of poles, both like and unlike poles must be tried. It is impossible to get the poles alone, so whole magnets must be used; and in order to have the poles free to move, one magnet must be suspended. The same device could be used as in Exercise 2, but a more convenient apparatus is a *compass,* which consists of a magnet supported on a sharp point and contained in a circular box, usually provided with a glass cover.

EXPERIMENT 1.

Apparatus.—Bar magnet; compass; wood; glass; paper; etc. (as in Ex. 1).

OBJECT.—To study what takes place when two magnetic poles are brought near each other, one pole being free to move.

MANIPULATION.—Take a compass and find the north pole (Ex. 2, Exp. 2). Lay the bar magnet on the table and bring the *north* pole of the compass-needle up to the *north* pole of the magnet. In the same way try bringing the *north* pole of the compass-needle up to the *south* pole of the magnet. Try also the *south* pole of the compass and the *south* pole of the magnet, and the *south* pole of the compass and the *north* pole of the magnet. Tabulate your results as follows:

Pole of Compass.	Pole of Magnet.	Results.

From the study of the table, write in your note-book the "law" of the action of magnetic poles.

EXPERIMENT 2.

OBJECT.—To see if the results of Experiment 1 still hold when various bodies are placed between the poles.

MANIPULATION.—Repeat Experiment 1 with pieces of wood, glass, paper, etc., held between the poles. Tabulate results as follows:

1st Pole.	2d Pole.	Body Used.	Results.

EXERCISE 4.

INDUCED MAGNETISM. BREAKING MAGNETS.

EXPERIMENT 1.

Apparatus.—Bar magnet; compass; large darning- or knitting-needle of quite hard steel; piece of iron wire; copper wire (say No. 16); large horse-shoe nail; iron-filings; piece of paper on which to put the filings; pieces of wood, and glass rod or tubing.

OBJECT.—To observe the effect of bringing a piece of steel in contact with a magnet.

MANIPULATION.—Take a large darning-needle and stroke one end, always in the same direction, several times on one pole of a bar magnet, as shown in Fig. 3. Note the nature of the pole used. Bring *the end of the needle that you stroked on the magnet* up to a compass-needle. Note what change has been produced in the darning-needle. Test the other end in the same way

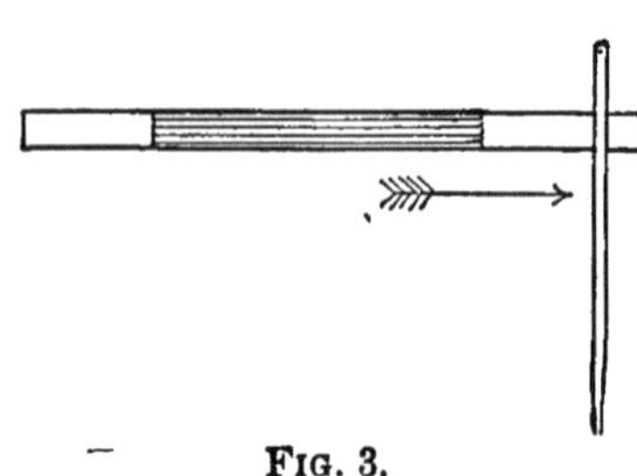

FIG. 3.

EXPERIMENT 2.

OBJECT.—To find out how the nature of the inducing pole affects the nature of the induced pole.

MANIPULATION.—Compare the nature of the pole induced in the end of the needle that was rubbed on the magnet with the nature of the pole on which it was rubbed. Test the nature of the poles by the knowledge gained in Ex. 3. Test also the other end of the needle. State in your note-book the law for the induction of magnetic poles.

EXPERIMENT 3.

OBJECT.—To see if all substances can be magnetized.

MANIPULATION.—Repeat Exp. 2 with the following: Soft iron wire which you are sure is not magnetized, wood, glass, and copper. Tabulate your results.

EXPERIMENT 4.

OBJECT.—To observe the results of breaking a magnet.

MANIPULATION.—Break your magnetized needle in the centre and, with the compass, examine both ends of each part. Record results, and draw figures illustrating them.

EXPERIMENT 5.

OBJECT.—To find what happens when the broken parts of the magnetare put together again.

MANIPULATION.—Bring the broken parts of Exp. 4 together, opposite poles in contact, and with the compass examine it all along its length for magnetism. From these results suggest, if you can, why the centre of a magnet shows no magnetism. Break one half of your needle and test each half as before. As the magnet is broken into smaller and smaller pieces, what will apparently be true of each part?

EXPERIMENT 6.

OBJECT.—To see if a change is produced in a piece of iron when brought near a magnet. (Compare this with Exercises 1 and 2.)

MANIPULATION.—Take a wrought-iron horseshoe-nail, test it to make sure that it is free from magnetism, and then hold it vertically with the lower end in some iron-fillings; bring the magnet over the upper end of the nail, but not in contact with it. (See Fig. 4.) Lift the magnet and nail together, *still holding them a little distance apart.* Observe results. Take away the magnet and again

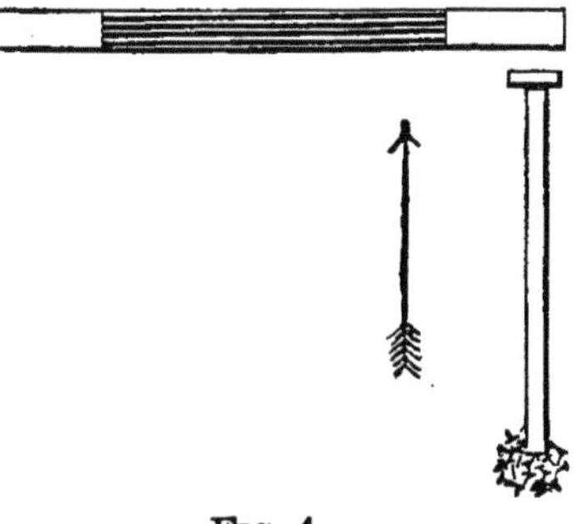

FIG. 4.

observe results. Repeat the experiment with the other pole of the magnet. Try putting a piece of paper between the magnet and the nail. Put the bar magnet entirely out of the way and test the nail for magnetism by means of the compass. After a lapse of five or ten minutes test the nail again.

State in your notes anything that you have observed regarding the different behavior of iron and steel when brought near or in contact with a magnet. If needed, Exp. 6 may be repeated roughly, with the nail and the magnet in contact.

QUESTIONS.—What tests could you apply to distinguish a magnet from a piece of steel? Why are magnets made of steel instead of iron?

EXERCISE 5.

LAW OF INDUCED MAGNETS.

Preliminary.—When a piece of iron or steel is magnetized by being brought near or in contact with a magnet, it is said to be an *induced magnet*, and the original magnet is said to be an *inducing magnet.*

The different behavior of iron and steel is expressed by saying that the steel has a greater *retentivity* than iron.

The magnetism remaining in a piece of soft iron after the inducing magnet has been removed is called *residual* magnetism.

In the following exercise, we wish to find out how the poles of the inducing magnet affect the poles of the induced magnet.

EXPERIMENT 1.

Apparatus.—Bar magnet; compass; 2 horseshoe-nails, or 2 pieces of soft iron wire, 3 or 4 inches long. If nails are used, they will probably have to be new ones, as the nail used in Exercise 4 will probably retain some magnetism.

OBJECT.—To study the nature of the poles of the *induced*

magnet, and discover how they are related to the poles of the *inducing* magnet.

MANIPULATION.—Arrange apparatus as in Fig. 5, where *AB* represents the bar magnet, *CD* represents a soft iron

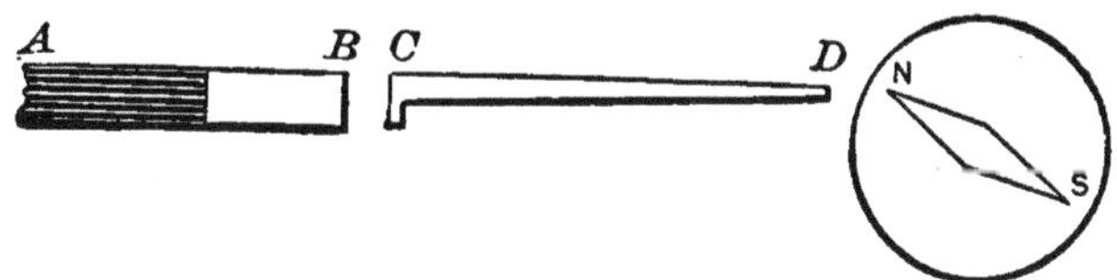

FIG. 5.

nail (free from all magnetism), and *NS* represents the compass. Be sure that the direction in which the nail and magnet lie is an east and west one, i.e., at right angles to the normal position of the compass-needle. Starting with the magnet three or four inches from the end of the nail, slowly bring it up to the end *C*. Note how the end D affects the north pole of the compass-needle. Repeat, using the south pole of the magnet. State in your notes how you find the pole of the induced magnet at the end furthest from the pole of the inducing magnet to compare with the inducing pole of the inducing magnet. Illustrate by a diagram.

EXPERIMENT 2.

OBJECT.—To discover the nature of the induced pole nearest to the inducing pole.

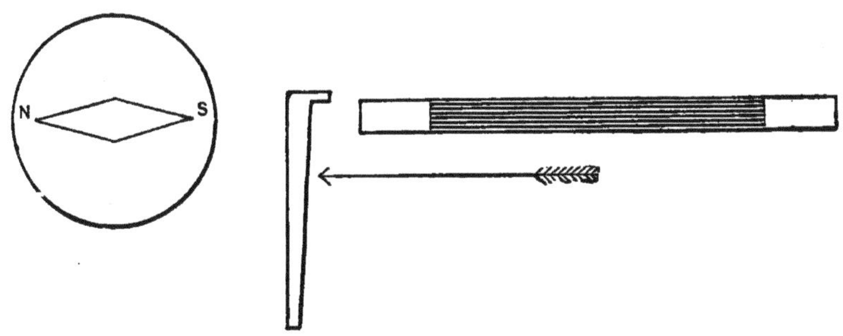

FIG. 6.

MANIPULATION.—Arrange apparatus as in Fig. 6. The distance from the pole of the magnet to the pole of the compass-needle should be about an inch. When the nail is brought close to the pole, it becomes an induced magnet and, if suddenly removed from the pole, retains its magnetism for an instant. (See Ex. 4, Exp. 6.) When the compass-needle is at rest, *suddenly* move the end of the nail which is in front of the magnetic pole up to the compass. The induced pole in the nail, remaining for an instant, is brought so much nearer to the pole of the compass that its action is noticeable, in spite of the greater action of the pole of the bar magnet.*

EXERCISE 6.

LINES OF MAGNETIC FORCE.

Preliminary.—We know from previous experiments that the space around the poles of a magnet is in such a condition that a body which can be magnetized is acted upon when brought into that space. This space is called the *field* of a magnet. If the body moves towards the magnet, it moves in the direction of the force, that is, in the direction in which the magnetic push or pull is exerted. Suppose a small piece of iron placed near a magnetic pole, as in Fig. 7. It is magnetized by induction and becomes a magnet. The end nearest the inducing magnet will be, as shown in the figure, a pole opposite to the nearest pole of the inducing magnet. The pole farthest from the inducing pole will be like it. The bit of iron will be attracted at one end and repelled at the other end with practically equal force, hence it will not tend to move toward the magnet, but will swing around until its length lies in the line of the push and pull.

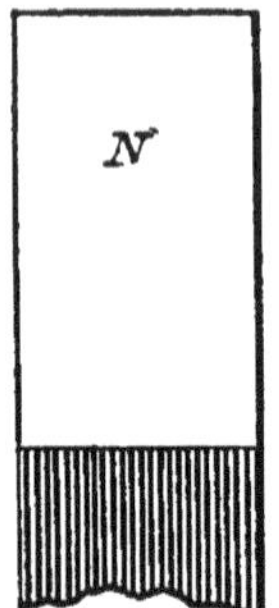

FIG. 7.

To illustrate this, imagine a stick lying on the floor,

---

* SUGGESTION.—Prepare an essay on Induced Magnetism.

with a string attached to each end, as in Fig. 8. On pulling both strings the stick will not move in either

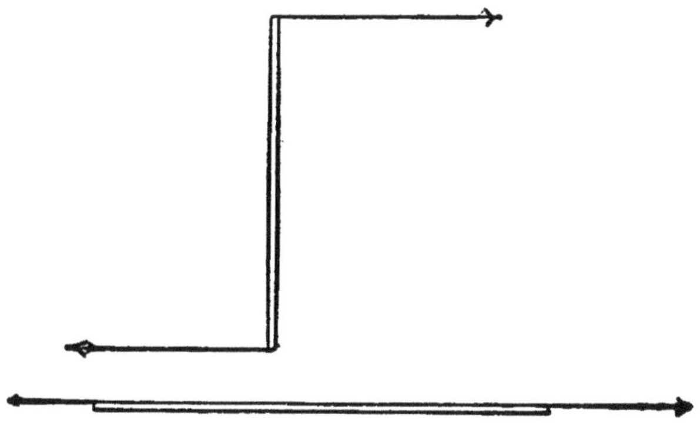

FIG. 8.

direction, but will twist around on its centre until it lies in the line in which the strings were pulled, as in the second diagram.

The lines marking the direction of the magnetic push or pull are called *lines of magnetic force.* In the following exercise we wish to observe these lines for various cases.

EXPERIMENT.

Apparatus.—A piece of sized writing-paper; fine iron-filings; two bar magnets; shellac; glass tube; blocks of wood.

OBJECT.—To study the lines of magnetic force.

MANIPULATION.—Take a piece of glazed paper, and pour fine iron-filings on it. Pour off the iron-filings and you will find that a layer of iron-dust remains on the paper. Lay a bar magnet on the table, hold the paper horizontally over it and then bring it straight down until it rests upon the magnet. If the magnet can be placed between two blocks of wood whose thickness is the same as the depth of the bar, so that the paper lies on a level surface, better results can be obtained. Tap the paper very gently with a lead-pencil and the little particles of iron will swing around on their centres until they everywhere lie in the line of the magnetic force. There will also be a number of lines on the paper which surround the outline of the magnet. These lines are called *lines of force,* and each line represents the line of the magnetic force in its part of the field. If a good set of lines is not obtained with one or two gentle taps, remove the paper and try again, as continual tapping will only cause the iron particles to bunch.

The positions taken by the iron-filings now give the diagram of the lines of force, and in order to study them they must be preserved in some way.

Method 1. Lift the paper *vertically* off the magnet to a height of about 6 inches, then gently place it on the table and carefully copy on another piece of paper the diagram obtained.

Method 2. With a lead-pencil trace out the different lines one after another on the paper itself. On wiping off the iron-dust you will have a fairly good reproduction of the lines.

Method 3. If it is desired to preserve in place the particles of iron themselves, there will be needed, in addition to the other apparatus, some thin shellac and a glass tube, open at both ends, about $\frac{1}{8}$ in. internal diameter and 3 in. long. Holding this tube as in Fig. 9, insert one end into the shellac and remove the finger from the top for an instant. Replace the finger and lift the tube from the liquid. So long as the finger is pressed on the top of the tube some shellac will remain inside and may be allowed to run out by raising the finger. With the paper lying upon the magnet, rest the tube very gently on the paper just inside the outline of the magnet (at the point marked x in Fig. 9), inclining it at an angle of about 45°. After it is on the paper, *but not before,* remove the finger, thus allowing the shellac to run out. The rate at which the shellac flows

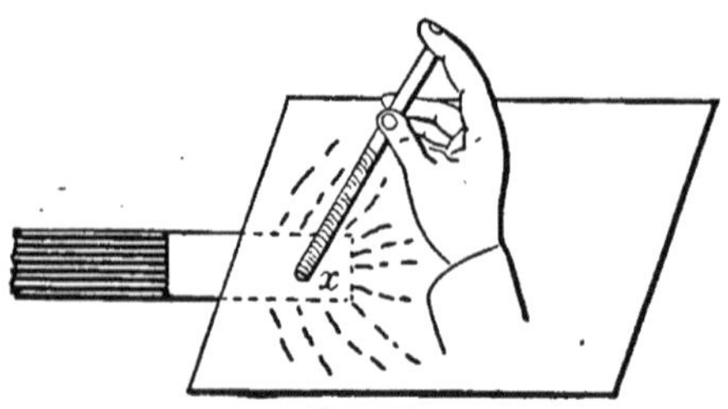

Fig. 9.

may be regulated by the degree to which the finger is removed. Care must be used to prevent the shellac from running out with sufficient force to move the iron particles. If the operation is properly conducted, the particles will not be disturbed and the shellac will soak in among them. If the first application of shellac does not entirely cover the figure, more may be added, with the same precaution, at other parts of the paper. The greatest care must be used not to allow any of the liquid to *drop* on the paper, as it will displace the iron-dust where it strikes. When the operation is completed, it is best to leave the paper until the alcohol has evaporated, which may be in five or ten minutes. During this time be careful that the paper is not disturbed. If in a hurry, the paper may be lifted *vertically* from the magnet and placed upon a horizontal surface until dry. When the shellac has become hard the iron is permanently fixed and the figure may be pasted in the note-book.

Method 4. For this the shellac should be somewhat thicker than in the preceding method. Place the paper over the magnet and pour the shellac on the paper, allowing it to spread out in a thin layer. Then carefully scatter the iron-dust over it and the iron particles will spread in the shellac, on which they will float. Tap gently until the lines appear, and allow the paper to remain undisturbed until dry.

By one of these methods obtain the lines of force in the following cases: 1. A north pole. 2. A south pole. 3. Two like poles. (Use two bar magnets end to end, the ends being about half an inch apart.) 4. Two unlike poles, arranged in the same way as in 3.

QUESTIONS.—1. What is the form of the lines of force around a single magnetic pole? 2. What is the shape of the lines of force around two like poles situated near each other? 3. What is the shape of the lines of force around two unlike poles situated near each other? 4. How does the form of the lines of force around two like poles compare with the form of the lines of force around two unlike poles? 5. In what case are the lines continuous from pole to pole, and in what case are they not? 6. How do the lines of force compare around a north or south pole when no other poles are near?

CURRENT ELECTRICITY.

EXERCISE 1.

VOLTAIC ELECTRICITY.

Introductory.—The word Electricity is the name given to the cause of certain phenomena, just as Magnetism is the name given to the cause of certain other phenomena, such as the attraction of iron and steel by a magnet or the attraction and repulsion of magnetic poles.

When chemical, as well as some other changes are produced under certain conditions, the resulting phenomena are said to be caused by *electricity.* When these phenomena are produced by chemical change, they are said to be due to current or *voltaic electricity.* In the following exercises some of these phenomena are observed, the conditions under which they can be brought about investigated, and some points connected with the useful applications of electricity examined.

EXPERIMENT 1.

Apparatus.—5 test-tubes ; a bit of zinc ; a copper tack ; an iron nail ; a piece of carbon (electric-light carbon or piece of old battery carbon); a piece of amalgamated sheet zinc; copper and zinc strips; diluted sulphuric acid ; compass ; tumbler ; some means of supporting the tubes in an upright position; bar magnet.

OBJECT.—To observe the action of dilute sulphuric acid on a number of substances.*

MANIPULATION.—Take 5 test-tubes and in each place 5 cc. 10% † sulphuric acid. Place in each tube one of

* Much time can be saved by working this experiment simultaneously with Experiment 2.

† One part of strong acid to nine parts of water.

the following substances: zinc, copper, iron, carbon, and a bit of zinc which has first been wet with acid and then rubbed with mercury.* From time to time during 15 or 20 minutes, watch carefully what goes on and record your observations. Watch particularly for answers to the following questions, but make in addition a complete record of all that takes place.

QUESTIONS.—1. Does any change go on? 2. If so, is it the same in all the tubes? 3. Has there been any change in the size of the bodies? 4. Arrange the names of the bodies in the order of the energy of the action, beginning with the body acted on the least and ending with that acted on the most.† 5. In any of the tubes is there any change in the nature of the action after it has gone on for a while? If such a case is noted, make a careful record of all that is observed in connection with it. 6. Summarize what has been learned regarding the action of cold dilute sulphuric acid on various substances.

EXPERIMENT 2.

Apparatus.—Strip of copper, strip of zinc (unamalgamated), with wires; a tumbler; dilute sulphuric acid; compass.

OBJECT.—To observe how electricity is produced from chemical action.

MANIPULATION.—You are provided with two strips, one of copper and one of zinc, a copper wire being attached to each. Place these strips in the tumbler and pour in 10% sulphuric acid until the plates are covered to a depth of about two inches. The acid must not be deep enough to cover the points where the wires are attached to the plates.

---

* Zinc so treated is said to be *amalgamated.*

† The bubbles of air which may slowly come out of some of the bodies, particularly the carbon, are not to be confounded with the bubbles of hydrogen gas given by the chemical action. The air has no smell, the hydrogen has. On thrusting a lighted match into hydrogen the gas will usually burn.

During this operation, do not allow the strips *to come in contact*. To prevent this, before pouring in the acid arrange as in Fig. 10, bending the wires back so as to spring against the sides, thus holding the two plates on opposite sides of the tumbler. Note carefully what happens on each plate, then bring the wires in contact, watching each plate carefully while you do so. Lay the compass on the table and arrange apparatus as in Fig. 11, so that when the compass-needle is at rest, one of the wires lies directly beneath and parallel to it. While arranging this, the wires must not be in contact. Now connect the ends of the wires.* What goes on in the tumbler? What goes on in the compass? Disconnect the wires and, when the compass-needle has come to rest, hold a bar magnet about six inches above it. Bring the magnet slowly down to the compass, watching the needle as you do so. Find in what direction the magnet must be held, and how it must be moved to produce the same effect on the compass-needle that was produced by (1) connecting the wires, (2) disconnecting them. How could a number of magnets be arranged so as to produce the same

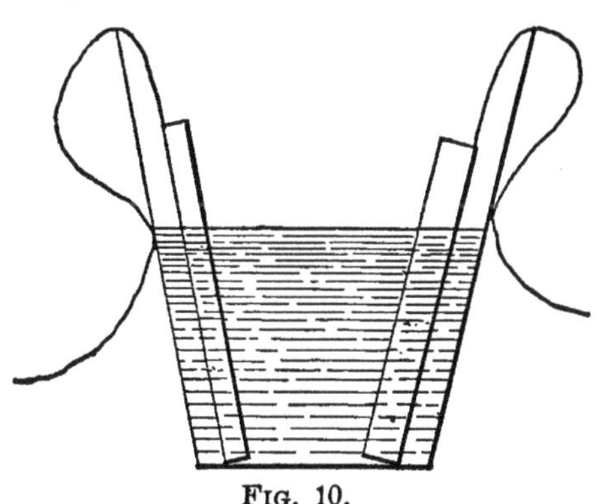

Fig. 10.

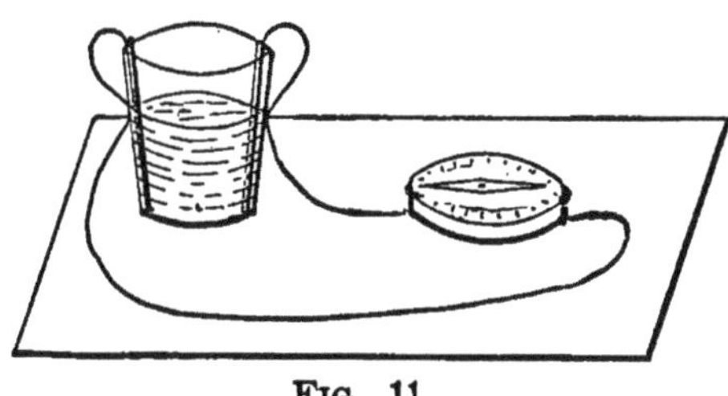

Fig. 11.

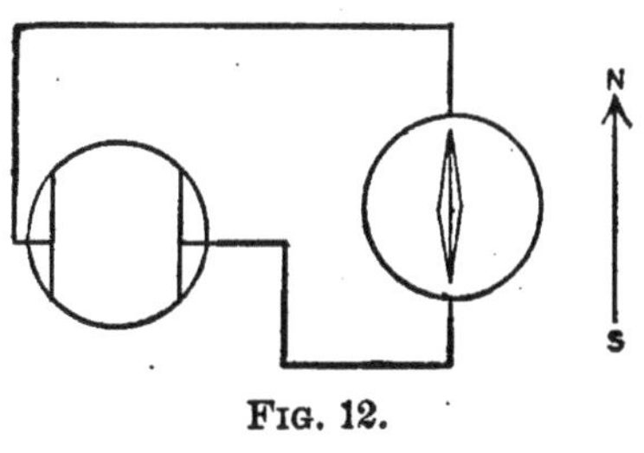

Fig. 12.

* Should any marked phenomena fail to appear on bringing the wires in contact, remove plates, clean carefully, and repeat.

effect on the compass as was produced by the wires? Illustrate by a sketch in your note-book.

EXERCISE 2.

CONDITIONS FOR PRODUCING CURRENT.

Preliminary.—When, on bringing a wire near a compass-needle, or any magnet free to deflect, a phenomenon like that in the preceding exercise is observed, the wire is said to have a *current of electricity* flowing through it. We do not know exactly what goes on in the wire, or why the needle acts as it does; and when we say that a wire is carrying a "current" we only mean that, if the wire were brought near a compass-needle, the needle would be affected as we have observed in the preceding exercise. The wire is the same wire whether it "carries a current" or not. When, for example, one plate is lifted from the liquid, the power of the wire to twist the needle disappears. This also happens if the wire is not complete from one plate to another. But while we do not change the wire, we do change the *conditions* under which the wire is placed. Hence when we speak of a wire carrying a current we imply a *special condition* of the wire. The steps necessary to develop in a wire the power indicated by the compass-needle are called the conditions needed for the production of an electric current. As will be seen later, there are other ways in which this condition of the wire may be brought about besides that of the galvanic cell; but no matter what has been done to a wire, if it can do what the wire in the preceding exercise could do, it is said to carry a current while in that condition, and at no other time.

QUESTIONS.—Define electricity; an electric current. How could you determine whether a telegraph-wire was carrying a current or not? Name any method for generating electricity besides the chemical one. Must anything be destroyed in order that the current may be generated?

Diagrams.—Figs. 11 and 12 both represent the set of apparatus used in this exercise. On examining them, however, you will see that Fig. 11 is a picture of the apparatus, showing it as it actually looks, while Fig. 12 bears no resemblance to it at all. In Fig. 12 the tumbler and plates are represented by a circle with two lines inside, the compass by a circle with a long diamond inside, and the wires by lines. It is not necessary to show what the tumbler looks like, or what the compass looks like, or what sort of wires are used; the essential thing is to make clear that the wire connecting the plates is carried under the compass in a north and south line, and this is done just as well in Fig. 12 as in Fig. 11. Such a figure as Fig. 12, which only shows the way in which the parts of the apparatus are arranged to bring about the conditions under which the experiment is worked, is called a *diagram;* and as diagrams are much easier to draw, in scientific work they are often used in place of pictures.

In making diagrams, each instrument has its own sign. Where a picture of an instrument is given in this book, the sign by which it is to be represented in diagrams is also given (for example, see Figs. 13 and 14). Unless otherwise instructed, always represent apparatus by diagrams in your note-book, and the instruments by the regular signs. Use a ruler whenever you can, and be careful to make the diagram large enough. A space at least 3 × 3 inches, and often even half a page or a whole page of your note-book should be used. A space reserved for a diagram should never have notes written in it. In a diagram the different parts are usually indicated by letters, generally the initials of the names of those parts; thus a compass is marked *C*, a wire *W*, etc. Where the same letters have to be used more than once, one or more accents are added. For example, if two wires were to be marked, they would be lettered *W* and *W′* (*W* prime), respectively, and a third wire would be marked *W″* (*W* second). Or capitals and small letters might be used.

EXPERIMENT 1.

Apparatus.—Copper strip; zinc of "tumbler cell" (amalgamated); tumbler; compass; dilute acid; iron plate; carbon plate of cell; water; nail; wood; glass; etc.

OBJECT.—(*a*) To observe the effects, in the tumbler and on the compass-needle, of amalgamating the zinc. (*b*) To study the conditions under which this effect on the compass-needle can be produced.

MANIPULATION.—*Part I.* Proceed as in the preceding exercise, noting carefully what goes on in the tumbler with amalgamated zinc, the wires first in contact, then not in contact. Repeat the test with the wire and compass.

QUESTIONS.—What effect has the amalgamation of the zinc on (1) the action in the tumbler when the wires are not in contact? (2) the action in the tumbler when the wires are in contact? (3) the action of the wire on the compass-needle?

Part II. (*a*) Bring the wires in contact and, holding one wire over the compass, cause the needle to deflect. Still holding the wire over the compass, separate the ends (it is well to tap the compass gently, as the needle is liable to stick). Having again caused the needle to deflect, raise one plate from the liquid. Note what happens in the compass and in the tumbler. Replace the plate, watching carefully for any changes. Try the other plate. (*b*) As regards the nature of the liquid: replace the acid in the tumbler by water, trying the compass test and watching the tumbler carefully. (*c*) As regards the relation of the plates to the liquid: place in the acid two strips of zinc, also try two strips of copper.* Repeat the compass-test. Try other metals—zinc and iron, zinc and carbon, iron and carbon. In each case try the compass-test and watch carefully what goes on in the tumbler.

QUESTIONS.—1. Can the compass-needle effect be produced with any two plates? 2. Do the plates all produce the same effect? If not, name them in the order of the amount of deflection they produce. 3. Is any change noticeable in the plates as the action goes on? If so, where? 4. Can the plates be connected by any substance? Lay a nail over the compass and touch the ends of the wires one to each end. Try in the same way wood, glass, wire, etc. Note the behavior of the compass in each case. (5) What conditions, then, must be fulfilled in order that the magnetic needle shall be deflected by the wire as regards (*a*) the wire, (*b*) the plates, (*c*) the liquid, (*d*) the magnetic needle?

Supplementary.—An arrangement of plates, liquid, etc., fulfilling the conditions found in the this exercise, is called

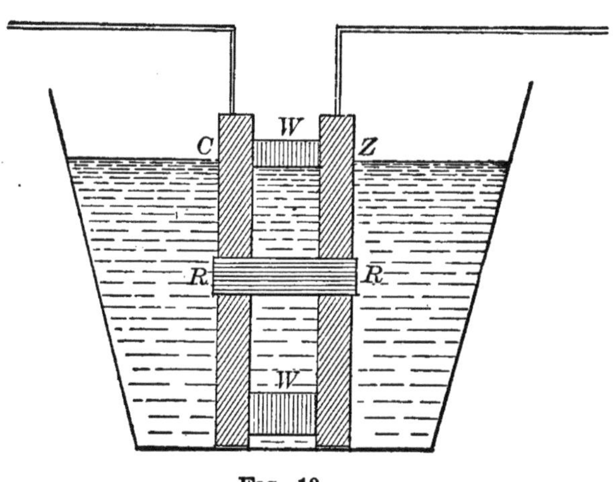

FIG. 13.

a *galvanic cell.* The plates are called the *elements* or *plates.* The fluid is called the *exciting fluid.* The

* Exchange strips with your neighbor.

wires leading from the plates are called the *conductors*, or the leading wires. One form of cell is shown in Fig. 13. The plates of carbon (*C*) and zinc (*Z*) are separated by pieces of wood (*WW*) and held in place by a rubber band (*RR*). The plates are set in a glass vessel to contain the exciting liquid, and to each plate is attached a wire, as shown in the figure. In diagrams a cell is usually indicated by two parallel lines of unequal length, as *B* in Fig. 14, which would represent one cell. The connecting wires are indicated by straight lines, as shown, while arrows near the lines indicate the direction of the current. When the cell is so arranged that the current is passing through the conductors, as in Fig. 14, the cell is sometimes said to be *running*. Although we only know that the wires possess certain properties when the cell is running that they do not possess when the cell is not running, it has been customary to imagine that electricity flows through the wire, and to speak of a *current of electricity* We have no means of finding out which way this current flows, but it is generally considered as flowing in the wire connecting the plates *from* the metal least acted upon *to* that most acted upon, and through the liquid in the cell in the reverse direction. The plate *from* which the current is supposed to flow in the wires is called the positive plate, and is indicated by the + sign. The plate *to* which the current flows is called the negative or minus plate, and is indicated by the – sign. The whole path of the current, plates, liquid, and conductors is called *the circuit*.

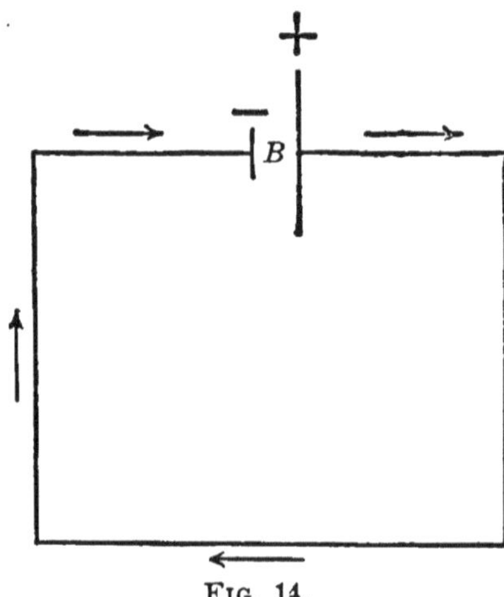

Fig. 14.

EXERCISE 3.

ACTION OF CURRENTS ON MAGNETS.

EXPERIMENT 1.

Apparatus.—About 100 cm. of No. 18 insulated copper wire; compass; electric current. Tumbler-cell. Supports for compass.

OBJECT.—To study the conditions affecting the behavior of a magnetic needle free to move, when near an electric current.

MANIPULATION.—1. Place the wire over the needle in a north and south line, arranging the wire so that the current flows from north to south. Complete the circuit and note the direction of the swing of the needle. 2. Repeat with the wire directly under the needle. 3. Repeat 1 with the current reversed, that is, flowing from south to north over the needle. 4. Repeat 2 with the current flowing from south to north. 5. Place the compass on a tumbler

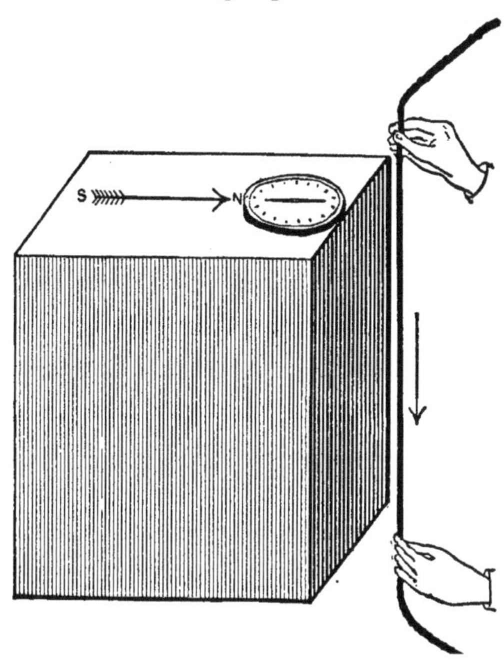

FIG. 15.

or block of wood as shown in Fig. 15. Hold the wire vertically, due north from the compass, with the current flowing down, and bring it slowly up to the north pole. 6. Repeat 5, holding the wire due south of the compass and bring it up to the south pole. 7. Invert the wire so that the current flows up, and repeat 5. 8. Repeat 6, the current flowing up. 9. Wind the wire tightly around the compass so as to form a rectangle. Hold this rectangle vertically in a north and south line; place the compass in the centre of the rectangle. 10. Repeat 9 with a loop instead of a rectangle. 11. Observe the effect of using more than one loop, by winding the wire around the compass once, then 5 or 6 times. 12. Observe the effect of the size of the loops. Try a large loop and a small one, keeping the compass in the centre of each. 13. Observe the effect of the distance of the wire from the needle.

QUESTIONS.—1. Illustrate each case by means of a diagram. 2. What conditions affect the direction of the swing of the needle? 3. If a man were swimming in the current so that it enters his feet and leaves his head, he always facing the needle, to which hand, right or left, would a north-seeking pole be urged?* A south-seeking pole? If you cannot tell, put yourself in the place of the imaginary man by holding the wire in front of you, or make a little paper man, marking the right and left hands, and, holding it in the position described, move it around the loop, observing towards which of his hands the north pole of the needle is deflected. In marking the hands of the paper man, do not forget that if he faces you his hands would be the opposite of yours, that is, his left hand would be opposite your right. 4. What conditions affect the amount of the swing? 5. Remembering that

---

* Notice that we are dealing here with *poles*, not with entire magnets.

the needle always tends to place itself in the line of magnetic force, from a study of the diagrams make a diagram of the field of force around a wire carrying a current.

The Galvanometer.—The law connecting the direction of the swing of the needle with the direction of the current is called *Ampere's law.* Advantage of this law is taken in constructing an instrument for observing the direction and strengths of electric currents. The instrument is called a *galvanometer,* and consists of a magnetic needle placed in the centre of a coil of wire and arranged so as to move freely. When this coil is placed in a north and south line and a current is passed through it, the needle is deflected. The direction of the deflection indicates the direction of the current, while the degree of the deflection indicates the relative strength of the current. In a general way, the stronger the current, the greater the deflection.

We found that with a given current the amount of deflection could be changed by varying the number of turns of wire. In most galvanometers this is done by changing the connections. With a very heavy current, but few turns would be needed; with a weak current, more turns would be required to give readable changes in the position of the needle for small differences in current. Of course, in comparing various currents, the same number of turns must be used.

Fig. 16 is a representation of a galvanometer. The coil of wire *W* is wound upon a wooden hoop, *H,* which is supported in an upright position by the base-boards *E* and *D.* This coil is connected with the three binding-posts *B, B′, B″.* Starting from *B,* the wire passes once around the hoop and is led out to *B′.* It then is wound around the hoop nine times more, so that if *B* and *B′* be connected, the current passes around the hoop once; if *B′* and *B″,* the current passes through nine turns; while if *B* and *B″*

be connected, ten turns are in circuit. A compass *C* placed in the centre of the hoop furnishes the magnetic needle whose indications are observed, the scale on the compass providing a means of measuring the amount of the

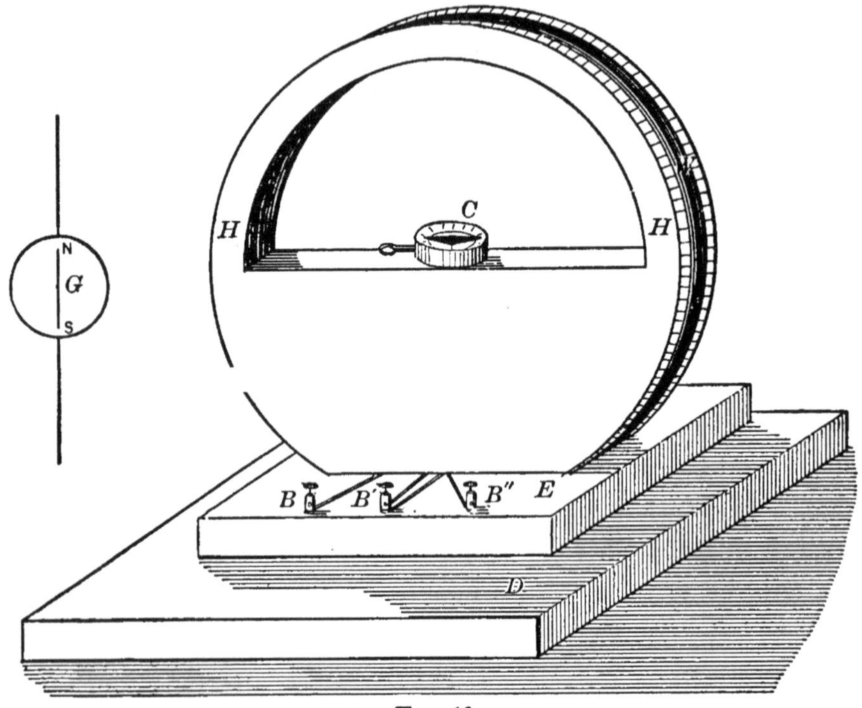

Fig. 16.

swing. A galvanometer is represented in diagram by the small figure on the left. This sign does not represent simply this form of galvanometer, but any form.

Precautions.—1. The compass must be in the centre of the hoop. 2. The coil must be in a north and south line. 3. There must be no iron or magnets near. 4. All contacts must be good. 5. Before reading tap the hoop gently, as the needle may stick. 6. In reading hold the eye as nearly as possible vertically over the needle. 7. The end of the needle when at rest should be directly over the zero of the scale. 8. When using the galvanometer connect it for ten turns, unless the instructions state otherwise.

To read by "Reversal."—As getting the zero of the scale just under the end of the needle requires quite nice adjustment and takes time, it is better to read the instrument by what is known as the "reversal method." Adjust the instrument, having the coil nearly north and south and the zero of the scale within five degrees or so of the position taken by the end of the needle when at rest. Close the circuit, read; reverse, and read again.* The average of the two readings will be the true deflection. This method saves time and is more accurate. It should be used whenever current strengths are to be compared.

EXERCISE 4.

CONDITIONS AFFECTING ELECTRICAL RESISTANCE.

Preliminary.—When a circuit is so arranged that the current can pass entirely through it, it is said to be "closed," and a circuit so arranged is called a *closed circuit.* When the current cannot pass at any point, the circuit is said to be "open," or "broken," and such a circuit is called a *broken* or *open* circuit. Connecting two points on a circuit so as to close it (*e.g.* bringing the ends of the wires together) is called *closing the circuit* or *making the circuit;* separating two parts of a closed circuit, so as to open it (separating the ends of the wires, lifting one plate from the liquid, etc.), is called *breaking the circuit* or *breaking contact.* Separating two parts of a circuit and attaching the ends to a conductor (as a wire or a piece of apparatus), so as to include it in the circuit, is called *introducing that body into the circuit.*

When several cells are arranged so as to give a current

---

* If no reverser is used, this may be done by changing the connections at the binding-posts.

they are called a *galvanic battery,* or simply a *battery.* That part of the circuit which connects the plates outside of the vessel is called the *external circuit,* and the part inside of the vessel is called the *internal circuit.*

We have already observed that when various bodies were inserted in the external circuit of the same cell, and the wires were laid over the compass at the same distance from the needle, the deflections of the needle were not the same. We have also learned that the amount of this deflection is, under similar conditions, taken as indicating the strength of the current. We naturally infer that all bodies do not possess to the same degree the property of allowing the current to pass. This idea is expressed by saying that the *relative conductivity,* or, more commonly, the *resistance,* of all bodies is not the same. A body that will transmit but little current is said to have a *high resistance;* one that will transmit considerable current is said to have a *low resistance.** Resistance, then, may be taken to indicate the degree to which a body possesses the property of *not* transmitting the current. In the following exercise we wish to find out—

1. What sort of bodies have high, and what sort low, resistance.

2. What effect *length, material,* and *cross-section* have on the resistance of bodies.

EXPERIMENT 1.

Apparatus.—For Part I: Cell; galvanometer; connecting wires; coil of copper wire; iron nail; pieces of zinc; carbon; wood; glass rod; dilute sulphuric acid and water; tumbler. For Part II: In addition, mercury cups; reverser; wire coils.

OBJECT.—To study (1) the power of various bodies to transmit the current, and (2) the conditions effecting this power.

---

* Of course, under the same conditions regarding cell, distance of wire from magnetic needle, etc.

MANIPULATION.—In Fig. 17, *B* represents a cell and *G* a galvanometer. One wire from the cell connects with the binding-post on the galvanometer. A wire leads from the galvanometer connected so that 10 galvanometer-turns are in circuit. If, now, the ends of wires $x\,x'$ are pressed on any substance, the circuit is completed. The ends of the wires should be scraped bright and free from insulation, and all contacts made perfect. The screws in the binding-posts should press firmly on the wires. These precautions are important. Set up the apparatus and show it to the instructor before beginning the experiment. The galvanometer must be carefully read according to the instructions already given.

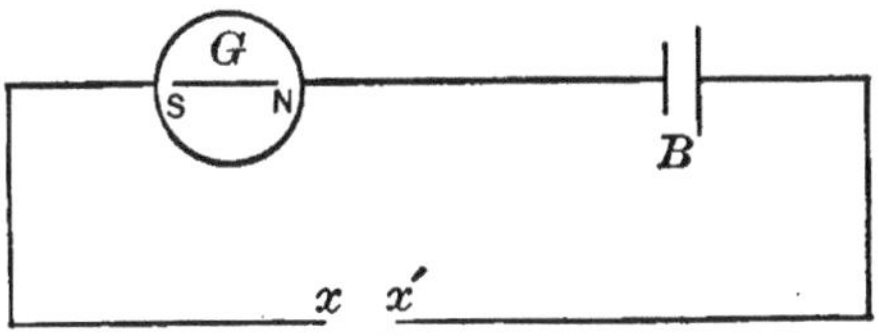

FIG. 17.

Part I.—Tabulate the results of inserting the following bodies, in turn, into the circuit: iron, copper, zinc, carbon, wood, glass, paper, water,* dilute sulphuric acid, etc., recording results as follows:

TABLE I.

Substance.	Galvanometer Reading.

Part II.—Introduce the mercury-cup, *M. C.*, into the circuit by attaching the ends to the binding-posts, one end to one post, as in Fig. 18. To introduce coils of wire into the circuit, dip one end well down into the mercury in one cup, and the other end into that in the other cup. Lay the card on which the coil is mounted flat on the table, and bend the ends of the wires so that they will not spring out of the mercury while the galvanometer is read. When

* To test the water, place the naked ends of the wires in a tumbler of water. Note and record anything going on in the tumbler, as well as the reading of the galvanometer. Then add to the water 10 or 12 drops of sulphuric acid, and repeat your observations.

two single ends, *a a*, Fig. 19, are placed in the cups, there are 10 meters of wire in circuit. When one single end, *a*, and one twisted end, *b*, are in the cups, there are 5 meters in the circuit. The coils are labelled, and con-

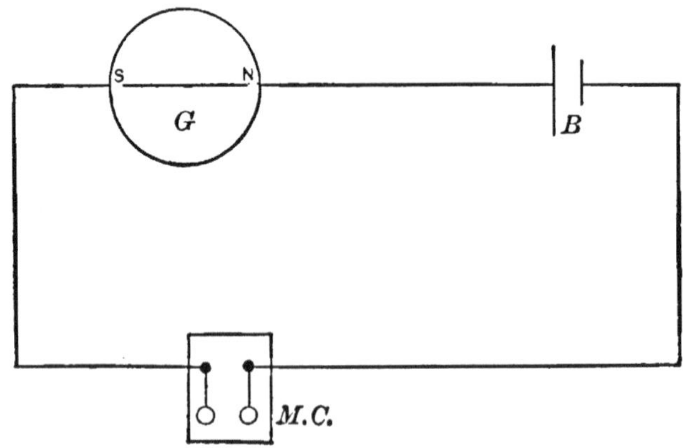

FIG. 18.

sist of, 1, 5 and 10 yards No. 28 copper wire ; 2, 20 yards No. 28 copper wire; 3, 5 yards No. 24 (larger cross-section) copper wire ; 4, 5 yards German-silver wire, No. 24 or No. 28, as labelled, same cross-section as one of the coils of copper wire of 5-meter length. Insert in turn 5, 10, 20

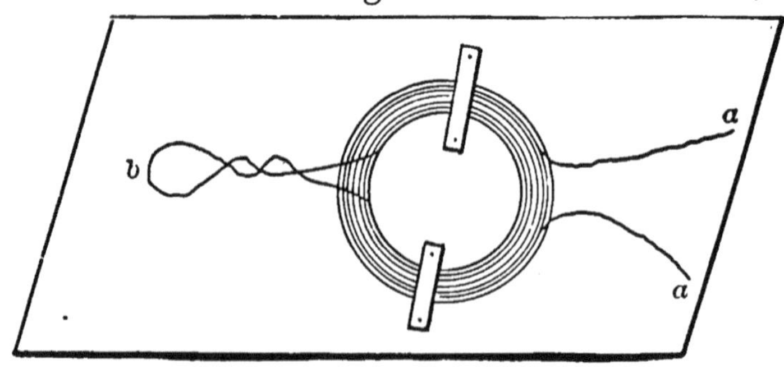

FIG. 19.

yards No. 28 copper wire, 5 yards No. 24 copper wire, 5 yards German-silver wire. Record results in :

TABLE II.

Material.	Length.	Cross-section. (By number.)	Galv. Read.

QUESTIONS.—Write in your note-books what has been learned as regards: 1. The equal resistance of bodies. 2. The sort of bodies that seem to have the highest resistance, and the sort that seem to have the lowest. 3. The resistance of water and the effect of adding acid. 4. The conditions affecting the resistance of a body. 5. The reason why the wires are of copper. 6. The reason why they are covered with cotton, etc. 7. The reason why the covering must be removed where connections are made. 8. Give a definition of resistance. 9. What part of the circuit has been used in this experiment? 10. Has anything been observed at the ends of the wires on making and breaking circuit? If so, what? 11. What goes on, so far as you have observed, when the current is passed through water? Or through water and acid? 12. Arrange the names of all the bodies that you have examined in a list, commencing with those of the highest and ending with those of the lowest resistance. Include air in this list.

EXERCISE 5.

ELECTRICAL RESISTANCE.

Preliminary.—If we wish to insert more than one conductor into the external circuit, connection can be made in two ways; either end to end, or side by side, as in Fig. 20.

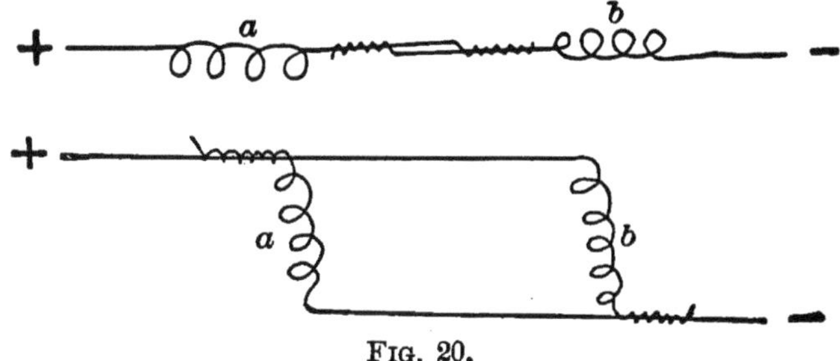

FIG. 20.

The first arrangement is called *connecting in series*, and the second *connecting in parallel*, or *in multiple arc*. The question is naturally suggested whether, with the same bodies, it makes any difference to the *total resistance* of the circuit which method is used. Write out the *general method* * of an experiment to answer this question, outline a *special method*, using the apparatus of the preceding exercise, and give the reasons for the use of each part of the apparatus. Prepare a diagram showing the relation of the parts, connections, etc.

EXPERIMENT 1.

Apparatus.—The same as in Ex. 4 (reverser) shown in Fig. 21, the mercury-cups being used to insert the various conductors in the circuit.

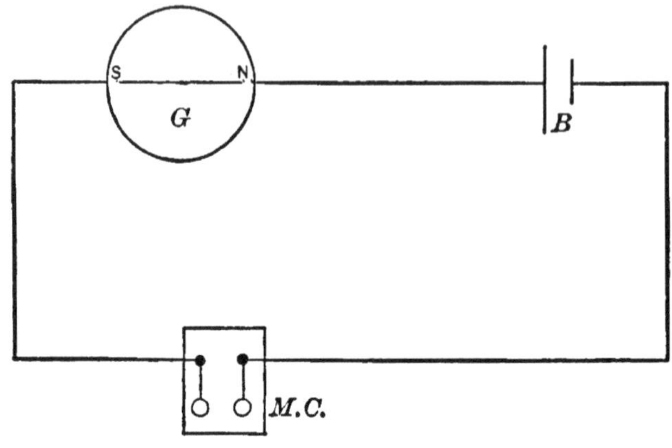

FIG. 21.

---

* The general method is a brief statement of the bodies experimented upon and of the conditions under which they must be placed in order to fulfil the object of the experiment. As different forms of the bodies may be used, and as the required conditions may be brought about by various means, the special method is a statement of just *what* forms of the bodies are used and just *how* these forms are placed under the conditions required by the experiment. The general method might be taken as the plan and the special method as the particular way in which the plan is carried out.

OBJECT.—To see if the manner in which conductors are placed in the external circuit affects the resistance of the circuit.

MANIPULATION.—*Part I.* Place in the circuit 20 meters No. 28 copper wire. Observe the deflection. Remove one end, attach to it, by twisting, one end of the 10-meter coil of No. 28 copper wire, and place the free end of the 10-meter coil in the mercury-cup. There are now in all 30 meters No. 28 wire in the circuit. Observe the deflection. Add to the other two coils the coil of No. 24 copper wire. Insert also the coil of German-silver wire. Record results as follows:

TABLE I.

No. of Trial.	Lengths and Substances Used.	Galvanometer Reading.		
		Right.	Left.	Average.

Part II. Place the coil of German-silver wire in the circuit and read the galvanometer. Without removing the first coil, also put in the circuit the 20 meters of copper wire. The current can now pass through both coils at the same time. Read the galvanometer. Also place in the circuit 5 meters No. 24 copper wire, in addition to the others. Record the results as above. Great care must be used in securing the contacts in the mercury-cups, and all precautions should be taken to get as accurate galvanometer-readings as possible.

QUESTIONS.—1. What effect is produced on the resistance of the external circuit by introducing conductors (*a*) in parallel? (*b*) In series? 2. Explain this result from the knowledge obtained in the preceding exercise. 3. Can the external circuit be composed of more than one sort of material?

EXERCISE 6.

METHODS OF CONNECTING GALVANIC CELLS.

Preliminary.—When more than one cell is to be used, there are two ways of connecting them ; as in Fig. 22, where the positive plate of one is connected with the negative plate of the next; or, as in Fig. 23, where all the positives are connected by one wire and all the negatives by the other. The former method is called *connecting in series;* the latter, *connecting in parallel,* or *multiple.* Now, the external resistance may be either comparatively high or low. For instance, the conductor may be a short, thick piece of copper wire, or a considerable length of fine German-silver wire. Hence it is a matter of some consequence to know which method of connection gives the most current, (*a*) when the external resistance is high, and (*b*) when it is low.

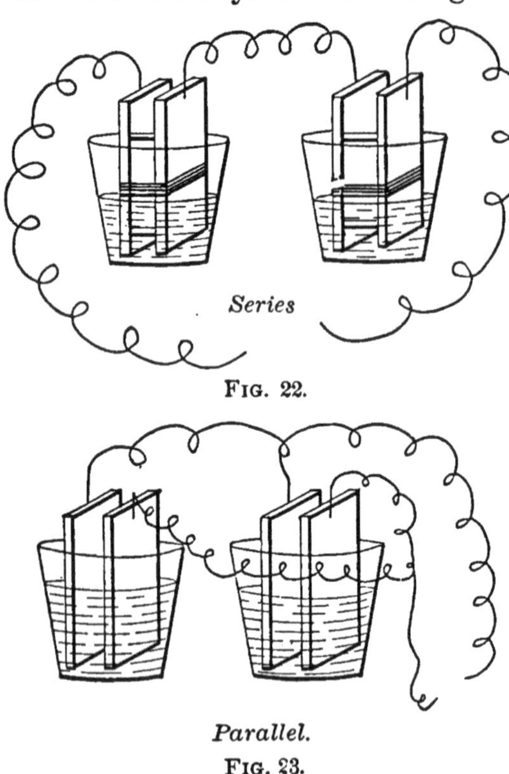

Fig. 22.

Parallel.

Fig. 23.

EXPERIMENT 1.

Apparatus.—Two cells; galvanometer; reverser; mercury-cups; short piece of thick copper wire; connecting wires (connecter, if available); wire-coil for high resistance.

Object.—To determine which method of connecting cells gives the most current, (*a*) when the external resistance is high, and (*b*) when it is low.

MANIPULATION.—Insert a galvanometer (one turn) and mercury-cups into the circuit of a single cell, as in Fig. 24. Connect the two mercury-cups with a short piece of copper wire and read the galvanometer. Disconnect the wire leading from the cell to the binding-post of the mercury-cup (marked *a* in Fig. 24), and by means of the connecter,

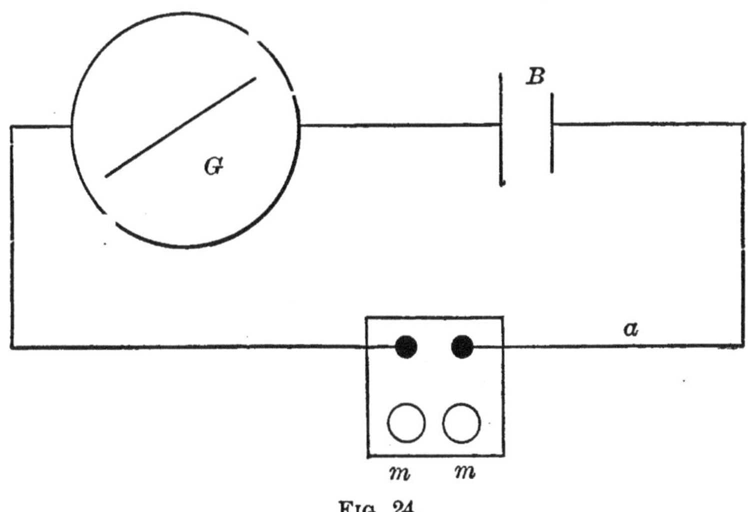

FIG. 24.

or by twisting the ends of the wires together, connect it with the wire leading from the opposite element of the second cell. Attach the other wire of the second cell to the empty binding-post of the mercury-cup. There are now two cells in circuit, the carbon of one connected with the zinc of the other—that is, the cells are in "series," as in Fig. 25. Connect the mercury-cups with the short piece of wire and read the galvanometer. The resistance of the external circuit is that of the connecting wires, the galvanometer connection, the mercury in the mercury-cup, and the short piece of wire connecting them; and as these are all good conductors, large and short, the total resistance of the external circuit is very small.

Take out the short piece of copper wire connecting the mercury-cups, and put in its place a coil of German-silver wire or 20 meters of No. 28 copper wire. This makes the resistance of the external circuit quite high. Change the galvanometer connections so as to have ten turns, and read the instrument. Leaving the coil of

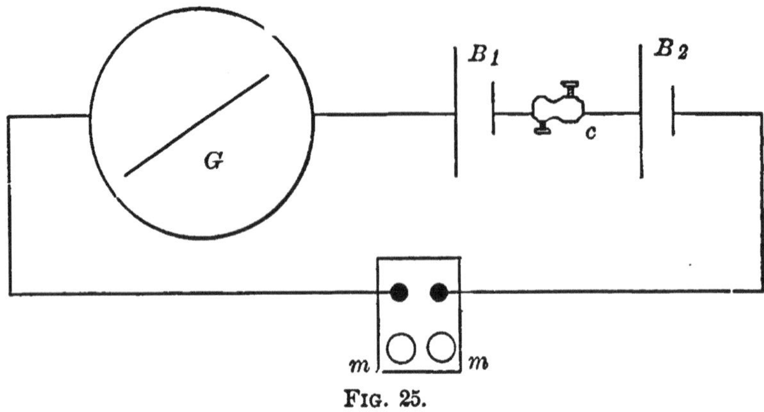

FIG. 25.

wire in the circuit, disconnect cell No. 2 (B_1) and connect cell No. 1 (B_2) alone with the mercury-cups. Read the galvanometer.

Connect the carbon plates of both cells with a wire leading to the galvanometer, as in Fig. 26. In the same way connect both zinc plates with the same binding-post on the

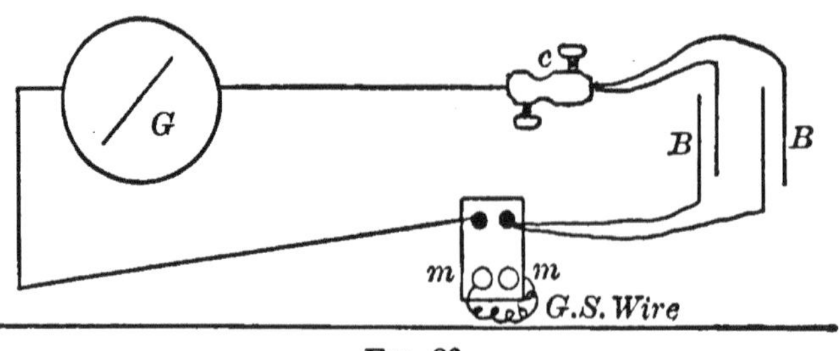

FIG. 26.

mercury-cup. The two cells are now connected in "parallel." Read the galvanometer. Replace the coil of wire in the mercury-cup by the short copper wire. Connect the galvanometer for one turn and read it. Arrange results as follows:

TABLE.

No. of Cells.	How Connected.	External Resist.	Galv-read.		
			R.	L.	Av.
1		High			
2	Series	"			
2	Parallel	"			
1		Low			
2	Series	"			
2	Parallel	"			

The cases in the table are not arranged as they are tried in the experiment. The table arrangement, however, makes it much easier to compare results. The table, made out in this way, should be placed in the note-book, and as the different cases are tried the galvanometer readings obtained can be placed where they belong in the last column. From the study of the table, which arrangement of cells seems to give the most current when the external resistance is high and which when the external resistance is low?

Unit of Resistance.—If resistances are to be compared accurately, a standard of resistance is needed. As we have found that resistance depends on length, cross-section, and material, we can get a standard of resistance by taking a specified length of a specified body of a specified cross-section. When the body used is mercury, cooled to the freezing-point of water, the length about 1 meter, and the cross-section 1 sq. millimeter, the resistance is said to be 1 ohm. Any body having a resistance equal to that of the standard mercury column is said to have a resistance of 1 ohm.

Resistance-box.—The instrument commonly used in measuring resistance is called a *resistance-box* or *rheostat.* It is an arrangement by which different coils of wire of known resistance may be placed in the circuit. A common form, together with a sign for a diagram, is shown in Fig. 27. It consists of a wooden box provided with three switches. A number of metallic heads are placed in semicircles about the pivots of the switches. Each head is furnished with a number which indicates the resistance

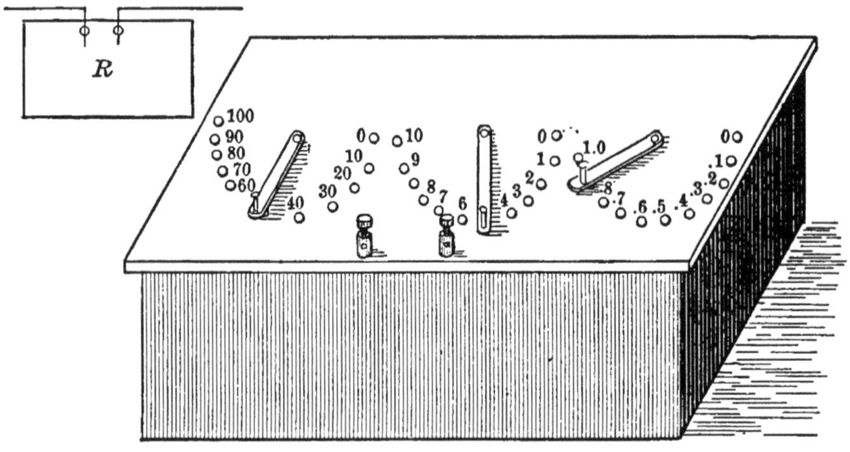

Fig. 27.

in the circuit when the end of the switch rests on that head. One switch (the one at the right in Fig. 27) places from 0 to 1 ohm in the circuit, the second from 1 to 10 ohms, and the third from 10 to 100 ohms. By using all three switches, any resistance from 0 to 111 ohms may be placed in the circuit. The sum of the readings of the three switches is taken as the measure of the resistance; for example, in the figure the box reads 55.9 ohms. The box has two binding-posts for making connections.

EXERCISE 7.

RELATIVE RESISTANCE.

Preliminary.—In the following experiment we wish to find the lengths of iron and copper wire equivalent in resistance to 1 centimeter of German-silver wire. We must place a known length of German-silver wire in the circuit, observe the deflection of the galvanometer, and then find the lengths of the wires of other materials required to produce the same deflection of the galvanometer. We will use the apparatus indicated in Fig. 28.

The board *A* carries two uprights, *BB*, which support a meter-stick, *M*. Stretched between these upright, is a piece of uncovered German-silver wire, on which slides a hook made of the end of one connecting-wire. Above the German-silver wire the iron and copper wires are stretched between the uprights, passing back and forth several times. A binding-post, *a*, is connected with one end of all three

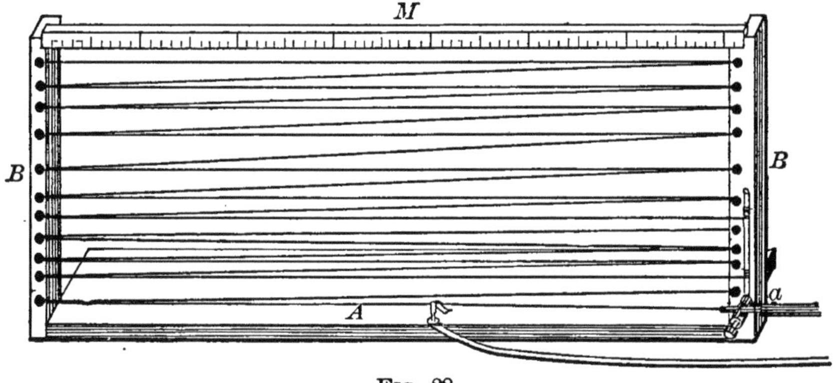

Fig. 28.

sets of wires. When the circuit is completed from the binding-post *a*, through the wire and the hook that slides on it, the length of wire in the circuit will be determined by the length of wire between *a* and the hook, as read on the meter-stick, and can be varied by sliding the hook along the wire, so varying the resistance. (Experiment 4, Part II.) More than one meter of the wire may be placed in circuit by attaching the hook to the wire running back to the right-hand upright. The total length of wire in the circuit in each case will be the *whole* length which the current passes through in going from the binding-post to the hook.

EXPERIMENT.

Apparatus.—Cell; rack with wires (Fig. 28); reverser; galvanometer; connecting-wires.*

Object.—To find what length of iron and copper wires give a resistance equivalent to that of one centimeter of German-silver wire.

Manipulation.—Having set up the cell and connected the galvanometer for ten turns, introduce the German-silver wire into the circuit, by attaching one connecting wire to the binding-post *a* and connecting the wire leading from the hook with the other element of the cell. The distance from the hook to the binding post, as read on the meter-stick, gives the length of German-silver wire in the circuit. Observe the galvanometer-reading with 50 cm. of wire in the circuit, then remove the hook from the German-silver wire, place it on the iron wire, and find by trial the position which gives the same galvanometer-reading. Repeat the operation with the copper wire, recording the total lengths in circuit in each case. Now, starting with 40 cm. of German-silver wire, again find corresponding lengths of iron and copper wire. Try some other lengths if time allows. Tabulate results as follows:

G. S. wire.	Galv.-read.			Iron wire.	Copper wire.	Iron wire equal to 1 cm. G.-S.	Copper wire equal to 1 cm. G.-S.
	R.	L.	Av.				

EXERCISE 8, A.

MEASUREMENT OF RESISTANCE.

Preliminary.—The purpose of the following exercise is to determine the resistance of a body by what is called the "method of substitution." This method is based upon

* One with hook, or an "English" binding-post.

the fact that bodies of equal resistance introduced into the same circuit will transmit the same amount of current. We find the resistance in ohms of a body which will transmit the same amount of current transmitted by a body of unknown resistance.* The apparatus used is as shown in Fig. 29. A galvanometer, *G*, a resistance-box, *R*, and mercury-cups, *mm*, are connected in the circuits of a

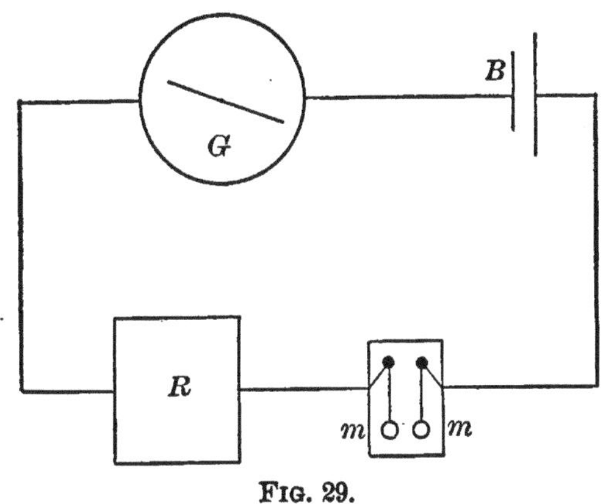

FIG. 29.

cell, *B*. The mercury-cups afford a means of easily inserting a body into the circuit. On replacing the body by a piece of thick copper wire the resistance of the mercury cups is practically reduced to 0, which is equivalent to taking them out of the circuit. By setting the switches of the resistance-box at 0, that also is practically taken out of the circuit, so that it is not necessary to disconnect it when the current is passed through the body.

EXPERIMENT.

Apparatus.—Cell; resistance-box; galvanometer; connecting-wire; conductor whose resistances are to be measured; mercury-cups; reverser.

OBJECT.—To determine the resistance of a body in ohms by the method of substitution.

* Let the pupil prepare a statement of the conditions necessary for carrying out such an experiment.

MANIPULATION.—Having all three switches of the resistance-box on the zero-point (that is, having no resistance in the box), connect the mercury-cups by each, in turn, of the coils of wire whose resistance is required, and read the galvanometer.* Replace the coil by the short piece of copper wire (so that the mercury-cups offer practically no resistance), bring your eye directly over the galvanometer-needle, and, without looking at the box, turn the 10-ohm switch until the galvanometer reads as close to the former reading as you can get, that is, until the addition of another 10 ohms makes it too low. Adjust now the ohm switch in the same manner, and lastly the 0.10-ohm switch. The adjustment of the switches should be done with the right hand, the eye being kept constantly on the needle. Read the resistance in the box. This is equal to the required resistance of the coil of wire. Repeat with the same coil, being careful not to look at the box so that your second trial may not be biased by the results of the first. Determine in this way the resistance of several coils of wire. Record as follows:

Length.	Material.	Galv.-reading.			Resistance.	No. of Wire.
		R.	L.	Av.		

* As all that is needed is to get the same galvanometer-reading, it is not necessary to reverse the galvanometer. In this case, however, all deflections must be on the same side. The first and second readings should be taken in as rapid succession as possible, in order to decrease the effect of any possible change in the current.

EXERCISE 8, B.

MEASUREMENT OF RESISTANCE.

Preliminary.—When, in such an apparatus as Fig. 30 represents, the current of the battery *B* comes to the point *a*, it divides, part going down the side *abd* and part down the side *acd*. If the resistances of the two sides are alike, the same amount of current will flow through each; but if they are not alike, more current will follow the side having the less resistance. If we attach wires at the points *b* and *c* and connect them with the galvanometer, so long as the same amount of current passes on each side the galvanometer will not be affected; but if the resistance of one side

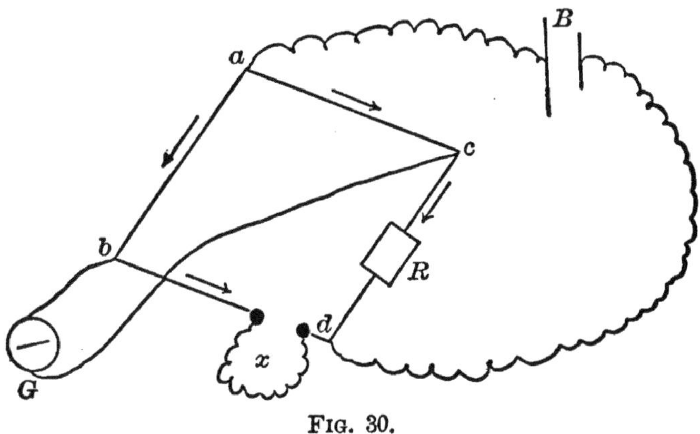

Fig. 30.

be made greater than that of the other, then the current will flow through the galvanometer from the side carrying the greater current. Suppose we insert in *cd* a resistance-box, and in *bd* the body whose resistance we wish to measure, *x*. If the resistance-box is at 0, the resistance of *x* will hold back the current in *bd*, and a portion of it will flow from *b* to *c* through the galvanometer, which will be deflected. Suppose now we increase the resistance in the box until the galvanometer reads 0, then we know that the resistance of *acd* equals the resistance of *abd* (since the wires of which the instrument are constructed have practically no resistance), and that the required resistance of x equals the known resistance in the box.

Wheatstone's Bridge.—On a board is placed a square of wire, *abcd*, Fig. 31, with a binding-post at each corner.

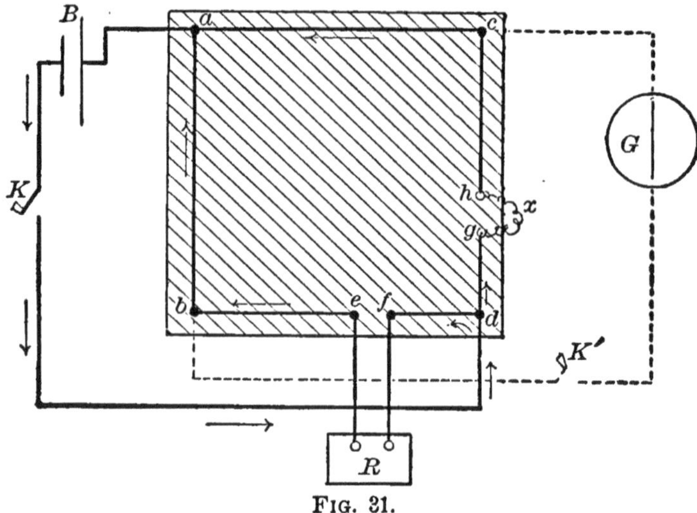

FIG. 31.

The sides *bd* and *dc* are cut, and two binding-posts inserted in each. From the binding-posts *b* and *c* wires (dotted lines) are led to the galvanometer *G*, and the battery-wires are attached to the binding-posts *a* and *d*. In the battery-

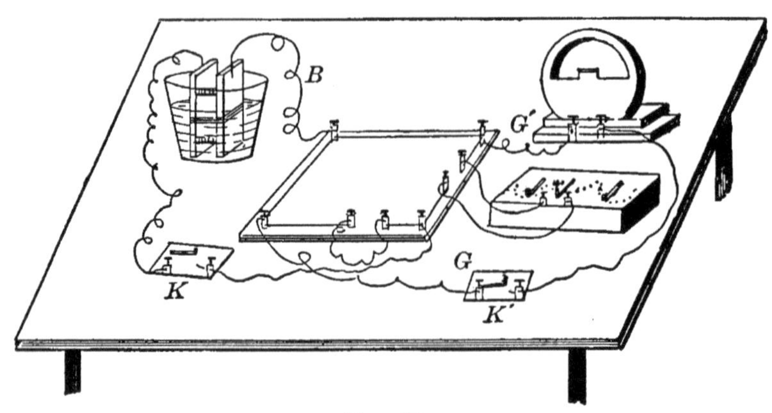

FIG. 32.

circuit is placed a key, *K*, which closes the circuit when depressed. A similar key, *K'*, is placed in the galvanometer-circuit. The substance to be tested, x, is attached to the binding-posts *gh*, and the resistance-box, *R*, is connected at *ef*. Fig. 32 is a *picture* of the same apparatus when ready for use.

EXPERIMENT.

Apparatus.—Wheatstone's Bridge; cell; galvanometer; two contact-keys; resistance-box; bodies whose resistances are to be measured; bar-magnet; connecting-wires.

OBJECT.—To determine the resistance of a body by the use of Wheatstone's Bridge.

MANIPULATION.—Connect the binding-posts *gh*, Fig. 31, by the substance whose resistance is required. Close the battery-circuit. Then close the galvanometer-circuit for an instant only, and observe the direction of the "throw" of the galvanometer-needle. Add resistances, observing the throw of the galvanometer-needle after each addition, until it begins to move the other way. Then work down with the resistance-box until, on closing both circuits, no tremble of the needle is observed. The resistance in the box is then equal to the unknown resistance.

PRECAUTIONS.—1. Avoid throwing the galvanometer-needle violently. This may be avoided by closing the galvanometer-circuit only for an instant until there are such resistances in the box that the galvanometer-needle scarcely moves. A longer contact is then safe. 2. Always close the battery-circuit before closing the galvanometer-circuit. 3. Be sure that all contacts are good.

During the first part of the experiment the motion of the needle may be very much deadened by placing a bar-magnet due north of the needle, in a north and south line, with its south pole near the north pole of the needle. During the latter part of the experiment the magnet may be removed. The presence of the magnet also tends to prevent the needle from oscillating, and saves time lost in waiting for the needle to come to rest after each trial.

EXERCISE 9.

ELECTRO-MOTIVE FORCE.

Preliminary.—We have observed that bodies differ from one another in resistance to an electric current, and have also noted the conditions upon which this resistance depends. Suppose, now, we place the same body in various circuits, will it always transmit the same amount of current? That is, have all currents the same power of overcoming a given resistance? Let us see if the conditions under which the current is generated make any difference. For this purpose try plates of various materials in the cell, being careful that they are the same size and distance apart, and that the external circuit is the same. With a galvanometer connected, we can see if the currents generated all have the same power of overcoming resistance.

EXPERIMENT.

Apparatus.—Tumbler-cell; plates of iron, lead, copper, with leading-wires; galvanometer; reverser.

OBJECT.—(1) To see if all currents have the same power of overcoming resistance. (2) To study the conditions affecting this power.

MANIPULATION.—Lay one of the plates on the table, put on it two little pieces of wood, lay the other plate on them, slip a rubber band around the whole, and place in the liquid in the tumbler. Connect the galvanometer, adjusted for ten turns, close the circuit, and read the galvanometer. If a reverser is available, use it; if not, reverse by changing connections. Try the cases given in the following table, and any others that you can.

The plates in each case being the same size and distance apart, and the external circuit the same, the resistance is practically alike in each trial. From the study of the re-

Positive Plate.	Negative Plate.	Galv.-reading.		
		R.	L.	Av.
Copper	Carbon			
Carbon	Zinc			
"	Iron			
Iron	Lead			
Carbon	"			
Lead	Zinc			
Carbon	Carbon			
Zinc	Zinc			
Copper	"			
"	Iron			
"	Lead			
"	Copper			

sults obtained, what do you infer regarding (1) the power of different currents to overcome resistance, and (2) the conditions affecting this power?

QUESTIONS.—Why are zinc and carbon usually used in galvanic cells? Would zinc and copper do as well? Why not use iron? What conditions seem to affect the results observed?

Supplementary.—The power possessed by a current of overcoming, or the power that pushes the current through the resistance, is said to be due to the *Electro-motive Force* (often written, for brevity, E. M. F.). The higher the E. M. F., the more current will pass through a given resistance, and the lower the E. M. F., the less current will pass through the same resistance. In a general way, the idea of electro-motive force corresponds to the "head" of water in a pipe; that is, to the pressure which pushes the water through the pipe. The higher the pressure, the more water will pass through the pipe; and the lower the pressure, the less will pass through the same pipe.

In the measurement of E. M. F., the standard is very nearly that given by a gravity-cell (zinc and copper in solutions of blue and white vitriol). The amount of current generated by the cell under test that will pass through a resistance is measured and compared with the amount that passes when a standard cell is used.

QUESTIONS.—1. Which gives the greater E. M. F., carbon and zinc, or iron and zinc? 2. What E. M. F. is produced by two like plates? 3. What seems to determine the E. M. F. in a galvanic cell?

EXERCISE 10.

ELECTRO-MAGNETISM.

EXPERIMENT 1.

Apparatus.—Carriage-bolt; 100 cm. No. 18 wire; cell; compass; iron-filings; paper and pencil; current; some means of varying resistance of the circuit and of reversing the current.

OBJECT.—To study the magnetic effects of an electric current and the conditions affecting the strength of an electro-magnet.

MANIPULATION.—*Part I.* Hold the carriage-bolt in front of you with the nut away from you and wind around it 15 or 20 turns of insulated wire from left to right (if a watch were facing you, the turns would be with the direction of the motion of the hands). Arrange as in Fig. 33.

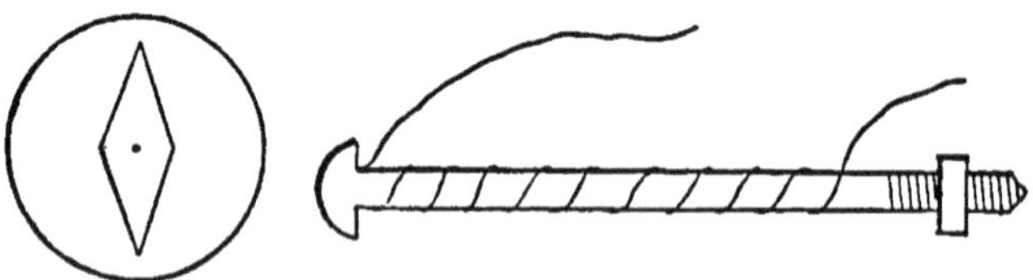

FIG. 33.

Complete the circuit. Note what happens. Test the other end at the compass. Move the compass along from one end of the bolt to the other. Note what happens to a piece of soft iron when a wire carrying a current is coiled around it.

Part II. Repeat Part I with the current reversed.

Part III. Rewind the wire in the opposite direction, repeat Part I, and observe the result.

Part IV. Reverse the current, and again repeat.

QUESTIONS.—1. What is the effect of reversing the current? 2. What effect has the direction of the current? 3. What happens when the circuit is broken? Does the effect produced by the direction of the current entirely disappear when the circuit is broken?

EXPERIMENT 2.

OBJECT.—To see if an electro-magnet* resembles an ordinary magnet.

MANIPULATION.—Bring the electro-magnet up to some iron-filings, and see if it attracts them.

EXPERIMENT 3.

OBJECT.—To investigate the conditions affecting the power of an electro-magnet.

MANIPULATION.—Lay a sheet of paper on the table, place the electro-magnet on it, and draw a pencil-line around the magnet so as to mark its position on the paper.† Place the compass about east of one pole of the electro-magnet, and at such a distance that when the current is turned on the magnet pulls the compass-needle around 15 or 20 degrees. Mark this position also. Change the current by varying the resistance in the circuit. Has the strength of the current any effect on the strength of the magnet? Put as many more turns on the magnet as you have already. Observe the results.‡ What effect has the number of turns? Give diagrams.

---

* A piece of iron magnetized as in Exp. 1, by an electric current, is called an *electro-magnet.*

† This precaution is taken in order that the distance from the magnet to the compass may be kept the same. Of course, for each test, both magnet and compass must be on the marks.

‡ In comparing the effects of different numbers of turns, the same current should be used.

QUESTIONS.—1. What must be done in order to obtain an electro-magnet? 2. What determines the nature of the poles? 3. What determines the strength of the magnet?

EXERCISE 11.

INDUCED CURRENTS.

Preliminary.—Arrange the apparatus as in Fig. 34,

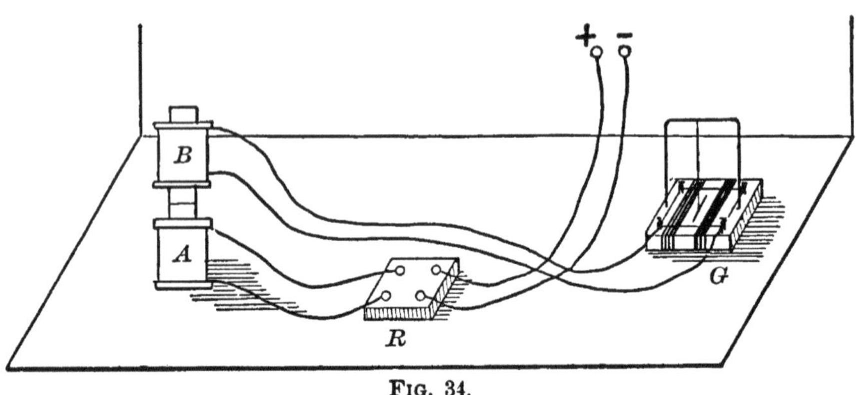

FIG. 34.

where the electro-magnet, *A*, is placed in the circuit, into which is also introduced the reverser, *R*. The ends of the coil of a second electro-magnet, *B*, are connected by flexible wires with the sensitive galvanometer, *G*.

When the current passes around *A* it becomes a magnet, and, if *B* is placed upon it, *B* becomes an induced magnet. Hence when the current is sent into *A* the result is equivalent to thrusting a magnet suddenly into the coil *B*, whose ends are connected with the galvanometer. By means of the reverser, the poles of *A* may be suddenly changed, and the same effect produced as if the magnet in *B* were suddenly withdrawn, turned end for end, and replaced in the coil. Slowly separate *A* and *B* and the effect is equivalent to slowly withdrawing the magnet from the coil *B*. Bring them slowly together, and the opposite effect is produced.

EXPERIMENT 1.

Apparatus.—Two electro-magnets; a strong current; connecting-wires; galvanometer; reverser; means of making and breaking the circuit, and, if possible, some means of varying current strength; clamp to suspend one coil.

OBJECT.—To observe the results of suddenly thrusting a magnet into a coil of wire.

MANIPULATION.—Place *B* upon *A*. When the galvanometer-needle is at rest, suddenly close the circuit of *A* (that is, suddenly introduce a magnet into *B*). Observe and record carefully the behavior of the needle.

EXPERIMENT 2.

OBJECT.—To observe the effect of withdrawing the magnet from the coil.

MANIPULATION.—While the current is passing through *A* and the galvanometer-needle is at rest, break the circuit of *A*. Observe and record as above.

EXPERIMENT 3.

OBJECT.—To investigate the effect of the nature of the pole used.

MANIPULATION.—Change the reverser and turn on the current in *A*. The poles of *A*, and hence those of the induced magnet in *B*, will be reversed. Observe results.

EXPERIMENT 4.

OBJECT.—To observe the effect of moving a coil of wire with an iron core in the field of a magnet.

MANIPULATION.—Remove the coil *B* from *A*. Close the circuit of *A*, and then move *B* up to and away from *A*, in various directions and with various speeds. Observe the galvanometer indications carefully in each case, with a view to stating the conditions affecting the amount of current, its direction, etc. If practicable, change the strength of the current through *A* to see if the strength of the magnetic field makes any difference. Tabulate results as follows:

How *B* was moved.	Direction of Galvanometer-reading.	Amount of Galv.-reading.

In the first column insert the words "towards *A*," "away from *A*," "in a horizontal circle above *A*," as the case may be. In the second column insert the words "right" or "left," as the case may be. In the third column, the words "more," "less," etc.

QUESTIONS.—1. Is it possible to obtain electricity by the use of magnets? 2. What conditions have you found to be necessary? 3. What conditions have you observed to affect the amount of the current,* and its direction? 4. Would it be possible to construct a machine on this principle which could be used to produce a current of electricity? 5. What would be the essential parts of such a machine?

A mechanical contrivance for fulfilling these conditions, and thus generating electricity from motion, is called a *Dynamo-electric Machine,* or simply a *Dynamo.*

EXPERIMENT 5.

OBJECT.—To see if, reversing the last process, motion can be obtained from a current.

MANIPULATION.—Set up the apparatus as in Fig. 35. Having *a* arranged as in Exp. 4, suspend *b* above it, and disconnecting the wires from the galvanometer, connect

---

* Notice that as the resistance of the galvanometer-coil *B* and the connecting-wires is always the same, any changes in the amount of current observed must be due to changes in the E. M. F. of the induced current; hence the conditions that seem to determine the amount of the current are really those which determine the E. M. F.

them with the circuit as shown, the reverser being in circuit with the lower magnet and *b* being suspended a little to one side of *a*.* Close the circuit through *b* and then through *a*, and notice the results. Suddenly reverse

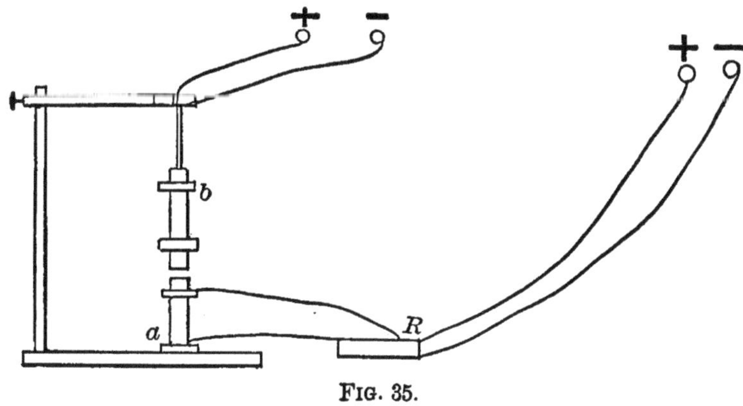

FIG. 35.

the current through *a*, and see if by reversing at the right time you can keep *b* swinging. A bar-magnet arranged as in Exp. 8 may be useful.

QUESTIONS.—1. Is it possible to obtain motion from an electric current? 2. Would it be possible to construct a machine on this principle which could be used to produce motion from an electric current? 3. What would be the essential parts of such a machine?

A mechanical device for fulfilling these conditions and thus obtaining motion from electricity is called an *Electric Motor*.

---

* Two cells may be used, or the two coils may be placed in parallel on one circuit.

MENSURATION.

NOTES ON MEASUREMENT.

Units of Measure.—If we wish to know just how many times one body is larger than another, that is, if we wish to make an exact comparison between the two bodies, we must ascertain the value of each one in terms of some fixed standard of measurement. Thus, to compare exactly the lengths of two boards, as a preliminary we refer both lengths to a fixed standard of length, and find out how many times this standard is contained in each of the lengths. Or, again, suppose we wish to compare the lengths of the two lines AB and CD, Fig. 36. Let us

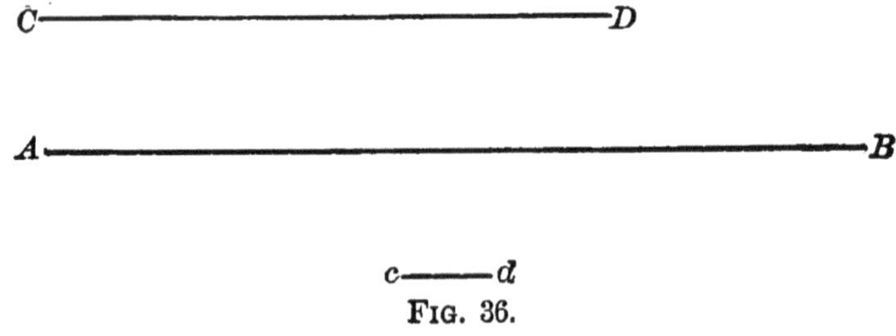

FIG. 36.

call the short line cd the standard. We find that AB is equal to 10 times cd and $CD = 7\ cd$. Then we know that the lengths of AB and CD are as 10 to 7. In case cd is not contained an integral number of times in AB and CD, it must be divided into known fractional parts, and the lengths expressed as so many cd's plus so many fractions of cd. In the same way, if we wish to compare volumes, a standard volume must be employed. Thus we can find how many times the little cube, *abcdef*, Fig. 37, is contained in the two larger cubes. If it is contained 36 times in one and 8 times in another, we say the volumes of the larger cubes are as 8 to 36. As a standard in comparing

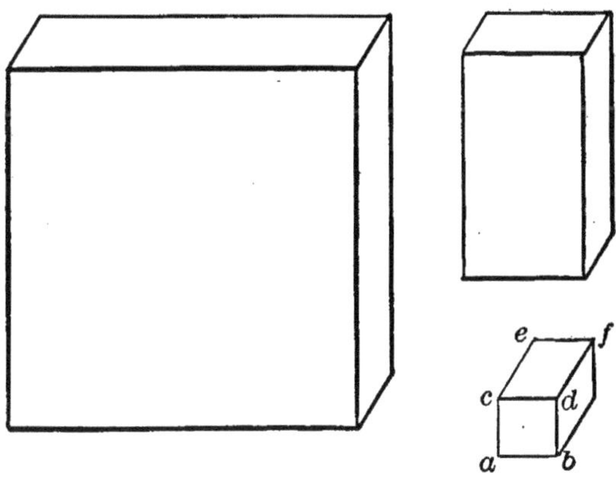

FIG. 37.

values we refer to dollars; in lengths, we refer to feet, meters, or miles; in volumes, we refer to quarts, gallons, or cubic inches, etc. Such a standard is called a *Unit.*

English and French Systems.—Of course any length or volume could be used as a unit, but in practice those adopted by the government or by custom are employed. Measurements of length, breadth, or thickness are called *Linear Measurements.* The common standard of length is the yard, which was originally adopted by the English Government as the length of a certain metal rod in their possession, and has been accepted by our own Government. The smaller units are the foot (one third of a yard), and the inch (one thirty-sixth of a yard, or one twelfth of a foot). These are called the English units, and in using them we are said to use *English Measure.* The inch is commonly divided into halves, fourths, eighths, sixteenths, etc. In scientific work, however, fractions of an inch are expressed as decimals. We also use the *French* or *Metric System.* Its standard is the length of a certain rod, called a Meter,

which is in the possession of the French Government. This is divided into 100 equal parts, called centimeters, and into 1000 parts, called millimeters. Thus a centimeter equals 10 millimeters. These prefixes, cent and mill, and their meanings, are familiar in our money.

Dollar.	Meter.	1.
Cent.	*Cent*imeter.	$\frac{1}{100}$.
Mill.	*Mill*imeter.	$\frac{1}{1000}$.

This system has the advantage of being divided decimally, so that a change from one unit to another can be effected by moving the decimal-point.

As a unit of Surface, we use a surface equal to that of a square measuring a unit of length on each side. Thus we have a square yard, square meter, etc. Such measure is called *Surface* or *Square Measure.* The surface expressed in square measure is the *Area.*

For comparing volumes, we use a volume equal to that of a cube, measuring a unit of length on each edge. Thus we have a cubic foot, cubic meter, etc. The unit used in our experimental work is generally the cubic centimeter, and sometimes 1000 cu. cm. or a liter. We have also quarts, gallons, etc., these quantities being the volumes of certain vessels belonging to the Government.

Change from one System to the Other.—If we know the value of one unit in terms of another, we can change from one system to another. The meter is 39.37 inches long. Suppose we wish to express 26 meters in yards. If one meter = 39.37 in., 26 meters = 26 × 39.37, or 1023.62 in., and as one yard = 36 in., $\frac{1023.62}{36}$ = number of yards = 28.85. Again, let us change 15 yards to meters. 15 yds. = 15 × 36 in. = 540 in. If one meter = 39.37 in., there will be as many meters in 540 in. as 39.37 is contained in 540, or 13.7 meters. Another method is as follows: One yard = 36 in., one meter = 39.37; then a yard $= \frac{36}{39.37}$ of a meter, and a meter $= \frac{39.37}{36}$ of a yard. If we change these fractions to decimals, we get 1 yard $= \frac{36}{39.37}$ meters, = 0.9144 meter; 1 meter $= \frac{39.37}{36} = 1.0936$ yds. Then if 1 meter = 1.0936 yards, 26 meters will equal 26 × 1.0936 yds., or 28.85. Similarly, if 1 yd. = 0.9144 meter, 15 yds. = 15 × 0.9144 = 13.7 meters.

Abbreviations.—Meter, m.; centimeter, cm.; millimeter, mm.; liter, l.; cubic centimeter, cu. cm.

It must be remembered that in surface measurement the units go by the square of 10, and in cubic measure by the cube of 10. Thus there are 100^2 or 10,000 sq. cm. in 1 sq. m., and 100^3 or 1,000,000 cu. cm. in a cu. m.

DETERMINATION OF LENGTH.

Scales.—The English Scale gives results in inches, and is shown as commonly arranged in Fig. 38. The long numbered lines mark inches; the shorter lines, half-inches; the still shorter, quarter-inches; and the shortest, eighths.

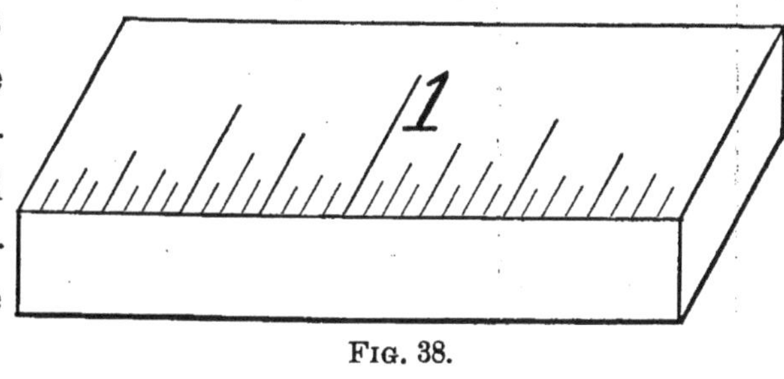

Fig. 38.

The results are here expressed in inches and decimals of an inch, not in vulgar fractions; as 0.125, instead of ⅛, or 2.375 in. for 2⅜ in. Yard-sticks are used in laboratories.*

The usual form of the Metric Scale is shown in Fig. 39. The long lines, as *aa*, mark decimeters; the shorter numbered lines, as *bb*, centimeters; the still shorter, *cc*, 0.5 cm., or 5 millimeters; the shortest lines, millimeters. Readings

* In some forms of meter-sticks there is an inch-scale on one side. Do not use these for *yard-sticks*.

are usually expressed in centimeters and decimals of a centimeter. For example, the scale to the line *AB* reads, in Fig. 39 (from right to left), 59 cm. plus 3 mm.; this would be recorded 59.30 cm. Meters are used for distances of over 100 cm. When the figures on scales are right side up, the scale reads from *right* to *left*. This must be remembered or mistakes will happen, as it is natural to try to read from left to right. Tenths of millimeters should be estimated by the eye.

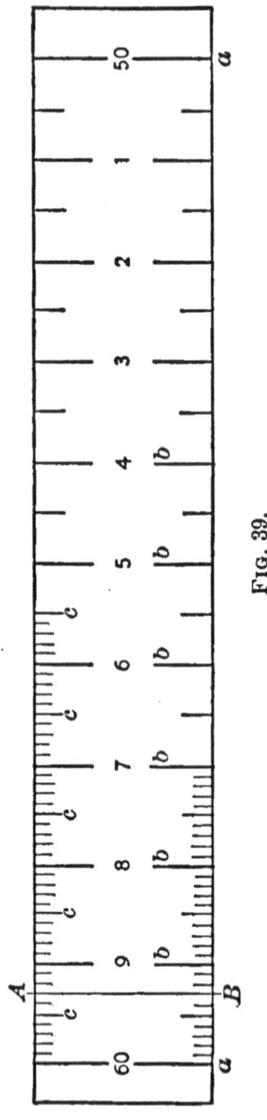

Fig. 39.

Reading Scales.—The scales are carried out to the edge of the rod (Figs. 40 and 41), and when possible this edge should be placed against the distance to be measured, as in Fig. 40, where it is desired to measure the distance between the lines *aa*. Where the scale cannot be applied directly to the object, be *sure that the line of sight is always perpendicular to the scale.* A metric scale should be read to $\frac{1}{10}$ mm., the fraction of a millimeter division being estimated by the eye. In general, an attempt should be made to read a scale to a fraction of the smallest division.

When determining shorter distances than the length of the measuring rod, where the body is loose and can be brought directly against the scale, it is not best to start at one end of the rod, but rather to bring the body to be measured near the middle of the rod, place one point on a numbered line (in a metric scale one of the decimeter lines), and read the position of each point on the scale. Suppose, for example, it is desired to measure the length of the box *ABCD* in Fig. 41. Lay the scale down on top as in the the diagram, and read the positions of *A* and *B*. Say $A = 26.14$ cm., $B = 42.81$ cm., then $B - A = 42.81 - 26.14$, or 16.67 cm.

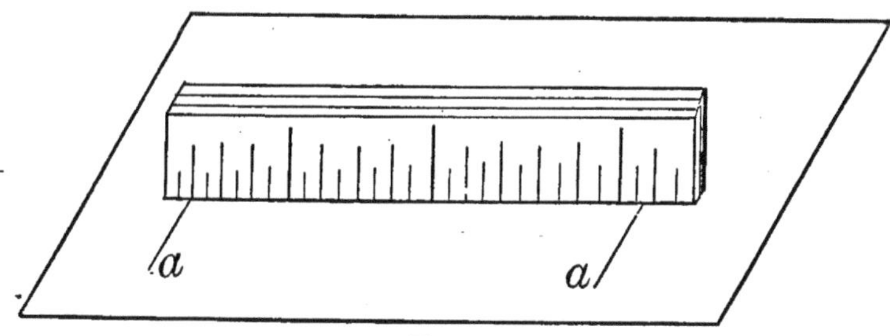

FIG. 40.

Suppose, again, the diameter of a sphere is needed. Obtain two rectangular blocks, *aa* in Fig. 42, and place the sphere between them, the whole resting on a level surface. Bring the blocks up against the sphere with their faces parallel. Then the distance between the blocks is the diameter of the sphere. Measure from edge to edge di-

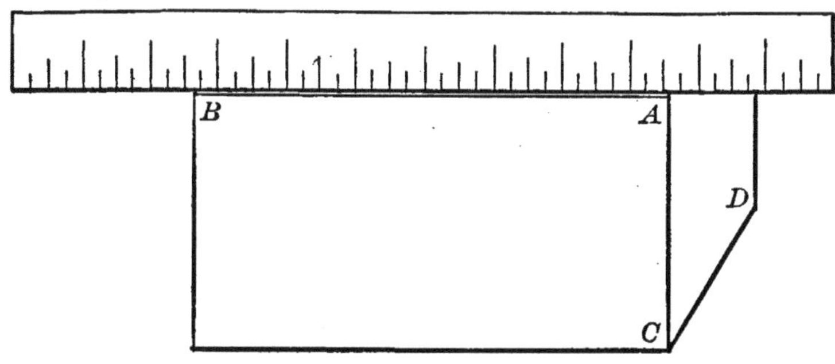

FIG. 41.

rectly over the centre of the sphere. The ball should be turned and the work repeated a few times. Both blocks should be set against a smooth vertical surface. In all these cases the same precaution is necessary,—to have the line of sight at right angles to the surface on which the scale is marked.

When the scale cannot be brought up to the body, special methods are resorted to. One is called The Compass Method. A pair of compasses or dividers are adjusted until the distance between their points is just that to be measured. The compasses are then applied to the scale, with one point on a numbered line, and the position of the other point is read.

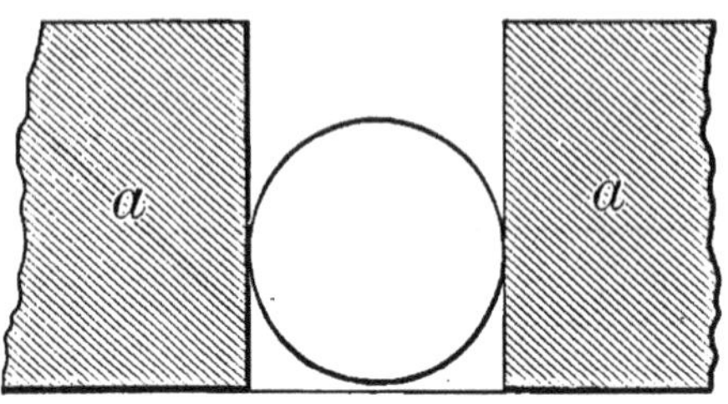

FIG. 42.

Sometimes a pointer is to be read against a scale, as in getting the Coefficient of Expansion. Always have the pointer as near the scale as possible, but not touching it, and arranged so that in moving it along the scale it is the same distance from it. Always read from the same side of the pointer. Best of all, end the pointer with a fine needle, and have the point of the needle read on the scale. A magnifying-glass is often of assistance.

EXERCISE 1.

PRACTICE IN THE USE OF LINEAR SCALES.

EXPERIMENT.

Apparatus.—A sheet of paper on which two crosses have been ruled with a sharp pencil, 30 or 40 cm. apart; a meter-stick with English scale on one side (or a meter-stick and a yard-stick).

OBJECT.—To determine: (*a*) The distance between the centres of the two crosses in inches. (*b*) The same distance in centimeters. (*c*) From these data,* the number of inches in a meter.

MANIPULATION.—Lay the sheet of paper smoothly upon the table, place the metric scale upon it, bring the centres of both crosses to the edge of the scale, and record their

* This word means the numerical results of the experiment, which are afterwards used in the calculation.

readings. Read the scale to 0.1 mm., observing all the precautions given in the "Notes on Measurement." Try five times, using different parts of the scale each time. Record results as in the following example, which gives the actual data of one set of determinations with this apparatus:

TABLE I.—METRIC MEASURE.

No. of Trial.	Reading L. H. Point.	Reading R. H. Point.	Length in Cm.
1	72.81 cm.	50.00 cm.	22.81
2	82.80 "	60.00 "	22.80
3	51.79 "	29.00 "	22.79
4	61.30 "	38.50 "	22.80
5	30.79 "	8.00 "	22.79

Average length in cm., 22.79 cm.

With the same precations, measure the distance between the centres of the crosses in the English system, reading to one-sixteenth (0.0625) inch, mentally changing the fractions of an inch to decimals. Arrange the results as follows:

TABLE II.—ENGLISH MEASURE.

No. of Trials.	Reading L. H. Point.	Reading R. H. Point.	Length in Inches.
1	1 inch.	9.937 inches.	8.937
2	5 inches.	13.975 "	8.975
3	7 "	15.959 "	8.959
4	10 "	18.968 "	8.968
5	13 "	21.975 "	8.975

Average length in inches, 8.963 inches.

CALCULATION.—Having now the values of the same distance in centimeters and in inches, the number of centimeters to the inch can be found by the following calculation:

Number of cms. : 100 cms. :: number of inches : x.

x = number of inches in 100 cms. (1 meter).

Taking the data given above :

$$22.79 \text{ cms.} : 100 \text{ cms.} :: 8.963 \text{ inches} : x$$

$$x = \frac{100 \times 8.963}{22.79} = 39.31,$$

which would be the value given by these data for the number of inches in the meter.*

EXERCISE 2.

THE RELATION OF CIRCUMFERENCE TO DIAMETER.

EXPERIMENT.

Apparatus.—Meter-stick (English and French scales), card-board circles, 10 cm., 20 cm., and 30 cm. in diameter; pencil; strip of paper or card-board about 1 m. long.

OBJECT.—To obtain answers to the following questions : (1) Is there any definite relation between the diameter of a circle and its circumference? (2) Do these results hold true for more than one method of measurement? (3) Is this relation the same for circles of different diameters? (4) Given the diameter of a circle, can the circumference be found? If so, how? (5) Given the circumference of a circle, can the diameter be found? If so, how? †

MANIPULATION.—*Method A.* Make a pencil-mark on the edge of the circular disk of paper (Fig. 43). Roll the disk along a straight line ruled on a piece of paper tacked to the table, starting with the pencil-mark just on the line, and rolling until the mark just touches the line again.

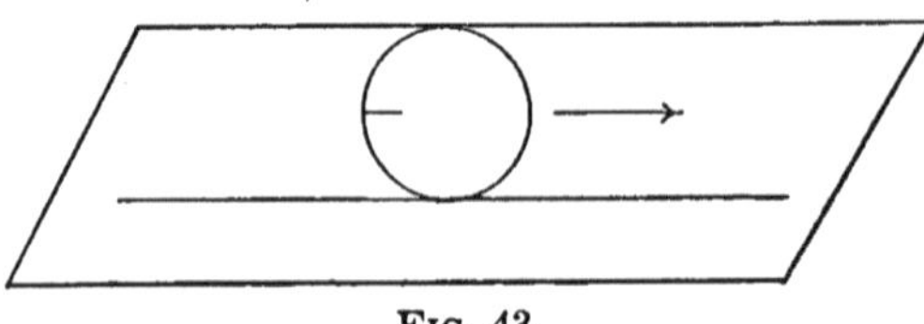

FIG. 43.

---

* These data give a value a little below the true one, viz., 39.37.

† As in all experimental work, first lay out a general method. In this case the general method is,—to answer question 1, measure the

Carefully measure the distance on the line between the two points touched by the pencil-mark. This gives you the circumference.

Method B. Lay the measuring-rod on the table and roll the disk along it, as in Fig. 44. This gives the required

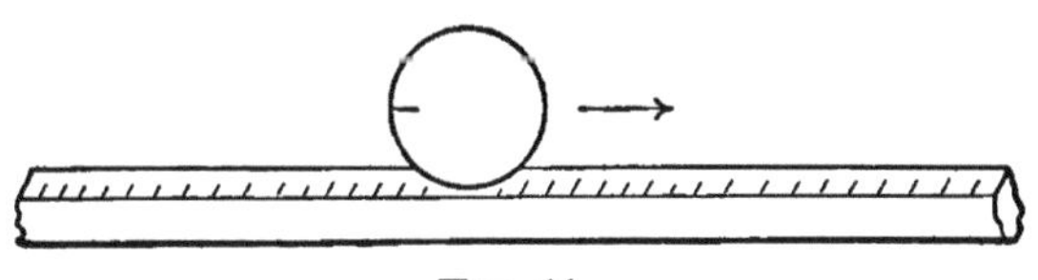

FIG. 44.

length directly on the rod. To measure the diameter, lay the disk flat on the table and measure the distance across it at its widest part. Make each measurement both in centimetres and inches; repeat three times, and record the results each time. The three numbers obtained for the circumference will not be just the same, owing to errors in manipulation, reading, etc. Add them, and divide the sum by the number of measurements. This gives the average measurement of the circumference.* Do the same for the diameter measurements. Then divide the average circumference by the average diameter. This gives a number which, except for errors, is that by which the diameter must be multiplied to obtain the circumference. It should be a little over three, and should be carried out to four decimal places. Do the same for the measurements in inches, and compare the numbers obtained in the two

---

circumference and diameter of a circle and find if any relation exists between them; to answer questions 4 and 5, use the knowledge obtained in answering question 1; to answer question 2, compare results obtained by using different methods of measurement; to answer question 3, use circles of different sizes.

* Of course this is to be done separately for the measurements in inches and in centimeters.

cases. Arrange in the note-book a table of results as follows :

CIRCLE I.

Measured in centimeters. Method "A."

Circum.	*Diam.*	Av. Circum.	Av. *Diam.*	Av. Circum. / Av. *Diam.*

CIRCLE I.

Measured in inches. Method "B."

Circum.	*Diam.*	Av. Circum.	Av. *Diam.*	Av. Circum. / Av. *Diam.*

Repeat in exactly the same way with Circle II. Answer in order the questions given above, and also state which method in your opinion yields the best results, and why?

CALCULATION.—There have now been obtained at least four values for the number which stands for the numerical relation between diameter and circumference. Average them, and the result should be nearly correct. Using this number, solve the following:

PROBLEMS.—1. A circle has a diameter of 6 cm.; find the circumference. 2. If the circumference is 25 cm., find the diameter. 3. If the diameter is 6 inches, find the circumference. 4. If the circumference is 25 inches, find the diameter. The number is represented in geometry by the sign π (pronounced "pī"). Then, with this sign, using D as the symbol for the diameter and C for the circumference, express as an algebraic equation the rule for solving such problems as the above.

DETERMINATION OF VOLUMES.

The Graduate Cylinder.—There are two kinds of measuring vessels in use for fluids—those measuring only a fixed quantity and those measuring varying quantities. Of the latter kind, the two commonly used in laboratory work are graduated cylinders and burettes. The graduated cylinder, often called simply a *graduate*, is a glass cylinder furnished with a foot so as to stand upright, and with a scale for reading volumes, usually in cubic centimeters. The scale is engraved on the glass, and arranged as in Fig. 45. There is a long numbered line for every 10 cu. cm., as *a a*. Between these are shorter lines, marking cubic centimeters, as *c c*; and between these, in turn, still shorter ones, marking half cu. cm. Such a scale is said to read to 0.5 c.c. There are usually two scales, one having the 0 at the top and the

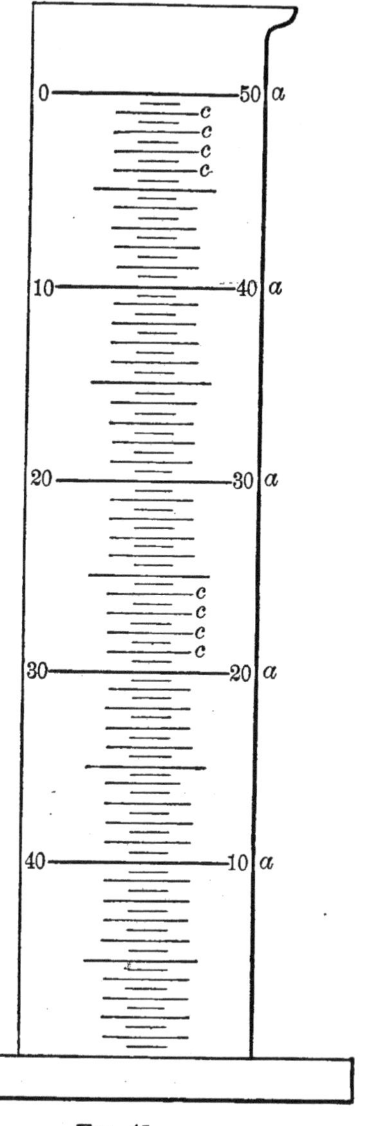

FIG. 45.

other at the bottom, the former used in measuring the volumes of liquids poured out of the cylinder, the latter in measuring the volumes of liquids poured in.

On looking horizontally at a graduate containing a liquid, the surface of the liquid appears as a dark band,

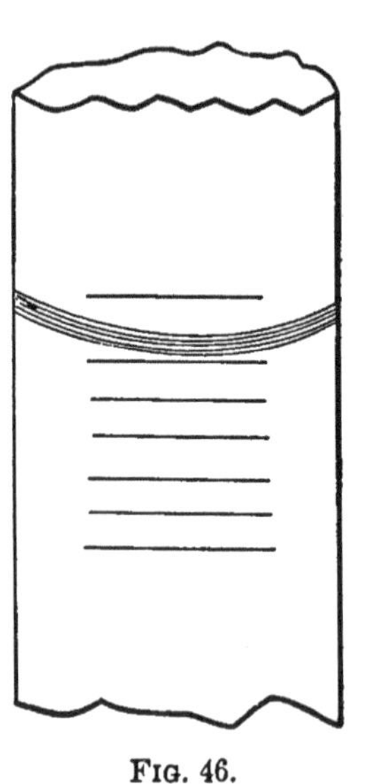

FIG. 46.

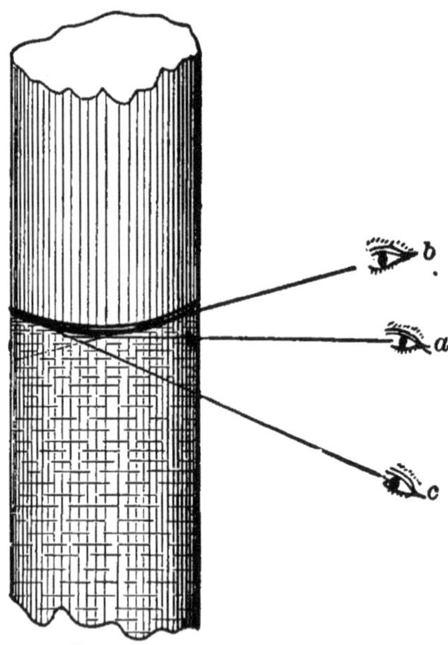

FIG. 47.

usually curving down in the centre, as in Fig. 46. This curve is called the *meniscus,* and the line corresponding to the *lowest* point of the meniscus is taken as the level of the liquid. The cylinder must be placed on a firm horizontal surface, and the eye brought to the level of the meniscus and directly opposite the scale. In Fig. 47 *a* represents the correct method of reading, *b* and *c* the incorrect. *b* would give a reading greater and *c* less than the true one. Errors caused by failure to read a scale with the line of sight in the proper position are said to be due to *Parallax.* The following method will avoid such errors:

Take a piece of stiff paper about $1\frac{1}{2} \times 4$ inches, being sure that the upper edge is clean and straight. Wrap it around the cylinder, and allow a portion to project, as shown in Fig. 48. Grasp the projecting portion with the finger and thumb, taking care that the upper and lower edges are even with each other. Draw the paper tightly around the cylinder, and move it until, on sighting across the top edge from one side to the other, the meniscus just

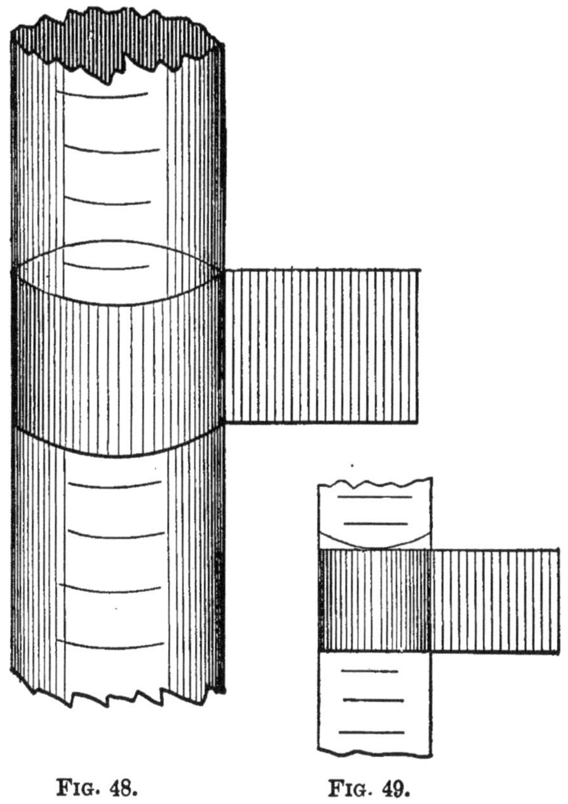

FIG. 48. FIG. 49.

touches it; Fig. 49. In this position the "sight" is horizontal. If the upper edge of the paper is just on a line of the scale, this gives the reading; but if not, the fraction must be estimated by the eye and expressed as a decimal. Suppose it comes, as in Fig. 50, between the 17 and 18 line. If half-way between, read 17.5; if one third the way, 17.33; if two thirds, 17.66; if three fourths, 17.75; etc. On the graduates commonly used remember that the space between two lines is *half* a cubic centimeter. Hence, if half-way, read 0.25; if one fourth above, 0.12 cu. cm.; etc. As a rule, any scale can be read to one tenth of its smallest division. When there is occasion to read mercury in a

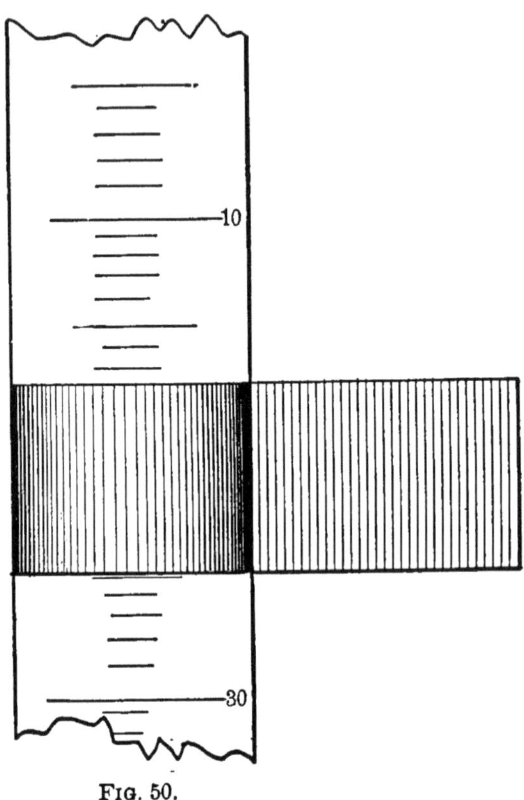

FIG. 50.

tube or graduate, it will be found that the meniscus is reversed (Fig. 51), curving up in the centre instead of down. In this case always read the level of the *top* of the meniscus, which can be done by the method just described.

When measuring out from a cylinder, it is generally best to fill it first to the 0 mark; though of course the level of the liquid can be read anywhere on the scale, and then read again after pouring off; the difference in the readings giving the volume poured off. In pouring from a cylinder, spilling can be avoided by holding one end of a glass rod against the lip of the jar, and the other end against the inside of the receiving vessel, as in Fig. 52. The liquid will follow the rod. On righting the cylinder

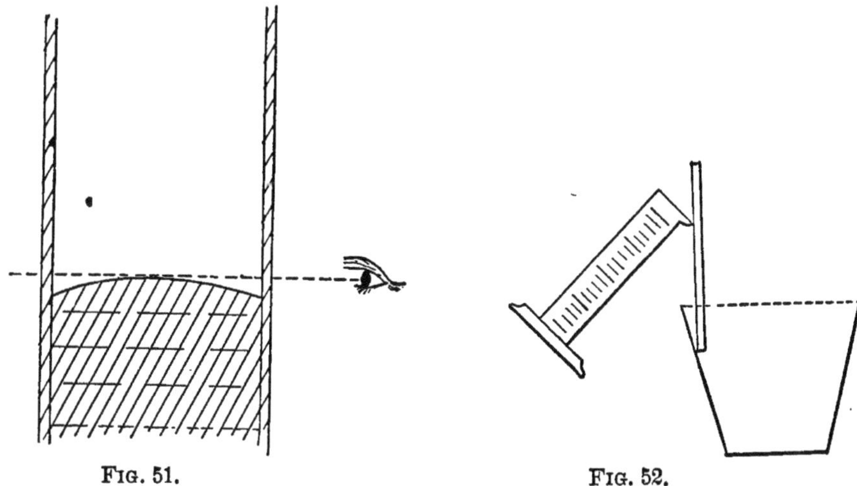

FIG. 51. FIG. 52.

always wait a moment before reading the volume, so that the water which has collected on the sides may run back again.

The Burette.—This is a long narrow tube having a scale on one side. (See Fig. 53.) A rubber tube is drawn over the lower end, which is narrowed, and the tube is provided with a spring clamp, *a*, which pinches the sides together. On pressing the ends of the clamp the rubber tube is allowed to expand and the liquid runs out at *C* at a rate proportioned to the pressure on the clamp. A burette is filled from the top and usually supported in a vertical position by a clamp. Burettes are usually graduated as in Fig. 54, which shows a portion of the scale. The long lines mark cu. cm. Each space between the cu. cm. lines is divided into five parts. Thus the scale reads to one fifth (0.2) cu. cm. The lines are numbered at every two cubic centimeters. In estimating fractions, one half a space = 0.1 cu. cm., one third = 0.06, two thirds = 0.12 cu. cm., etc. The scale may be read with a piece of paper like a graduated cylinder.

Many burettes are provided with a *float*, consisting of a bulb of glass shaped as in Fig. 55, with a little mercury in the lower end so that it will float upright. A line is ruled around it (*a* in Fig. 55). When the float is placed in the burette it floats in the liquid, rising and falling with it, and the line is observed on the scale through the glass. The position of the line is read as the level of the liquid. As the float is of nearly the same diameter as the inside of the burette (thus bringing the line very near the scale), with ordinary care there is little danger of error due to parallax. It is well to tap the burette gently before reading, as the float is apt to stick a little to its sides. When filling the burette, pour the liquid in until the line marking its level is above the 0 mark, which is near the top of the scale, and carefully draw out the excess until the reading is 0, allowing the waste liquid to run into a vessel provided for the purpose. To make sure that no air remains in the rubber tube, fill the burette part full, place the upper end in the mouth, and, holding the clamp open, blow the liquid nearly out. This repeated several times will generally drive out the air.

Instead of a clamp, burettes are sometimes provided with the arrangement shown in Fig. 56. A piece of glass rod about a fourth of an inch long, whose diameter is such that it will fit tightly in the rubber tube, is cut off, and the ends are rounded by heating. The little plug thus formed is thrust about half-way up the tube. By pinching with the forefinger and thumb on one side of the tube at the point where the plug is situated, a channel is formed through which the liquid can pass. The rate at which the liquid flows depends upon the amount of the pressure.

Graduated Flask.—This is a vessel having a mark engraved upon its neck. When filled so that the lower end of the meniscus just touches the mark, the flask holds a fixed quantity of liquid. Those commonly used hold 1

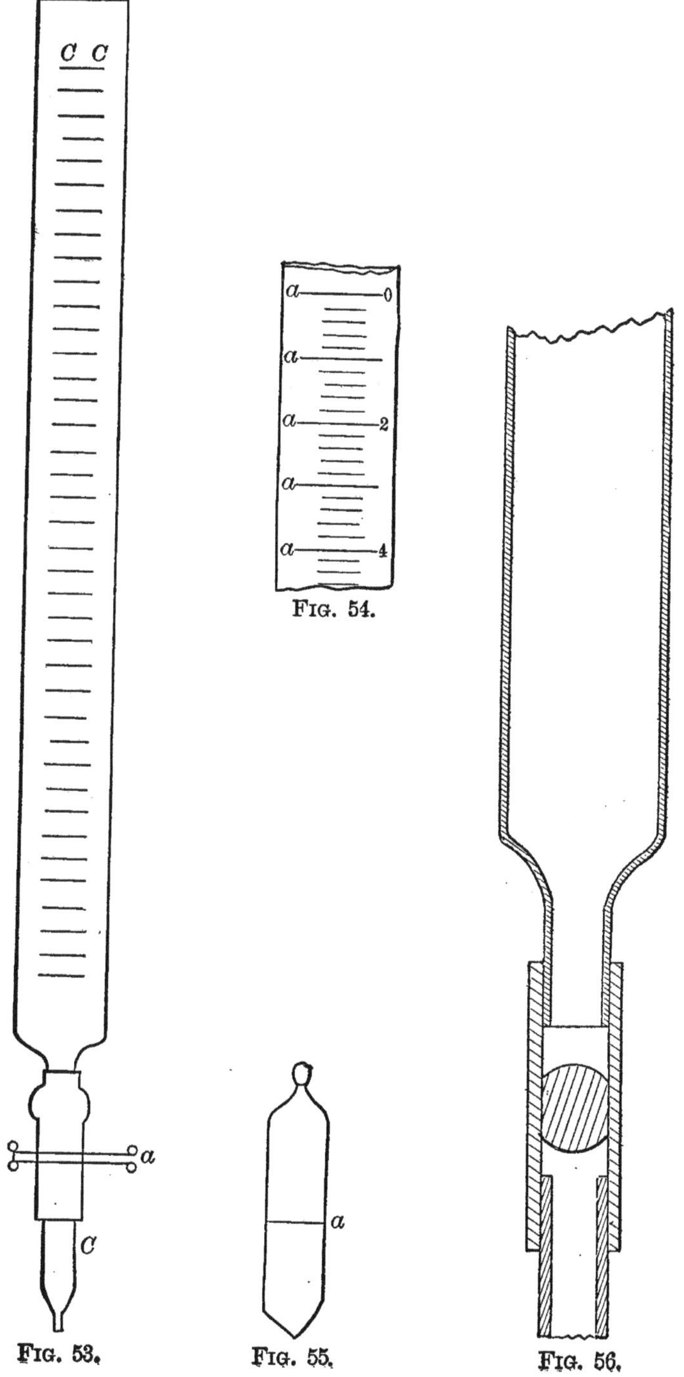

FIG. 53. FIG. 54. FIG. 55. FIG. 56.

liter (1000 cu. cm.), one-half liter (500 cu. cm.), and one-fourth liter (250 cu. cm.), their capacity being marked upon them. For accurate measurements, fill the flask nearly to the mark, and placing the eye so that the mark appears as a straight line, drop in the liquid with a piece of glass tube * until the correct level is reached. If you fill above the mark, use the tube to withdraw some of the liquid. The flask should rest on a firm horizontal surface. It is well to mark the position of the line by a piece of paper, as described in the directions for using graduates.

EXERCISE 3.

PRACTICE IN DETERMINING VOLUMES.

EXPERIMENT.

Apparatus.—Method A: Graduated cylinder; body whose volume is to be measured; water.

Method B: Piece of fine wire; cylinder; rubber band; paper for markers; body; water.

Method C: Burette or equivalent; ungraduated jar or equivalent; rubber band; paper for markers; body; water.

OBJECT.—To determine the volume of an irregular body by displacement.

MANIPULATION.—*Method A.* Fill the graduated cylinder part full of water, read the volume (as *a* in Fig. 57), observing all the precautions given in the preliminary notes on determination of volume, and record the reading. Drop the body whose volume is to be determined (as *e*) into the cylinder, and again read the volume (as *C*). The difference in the readings on the scale (*a* to *C*) is the volume of the body.

Method B. Attach a fine thread or wire to the body whose volume is to be determined, and drop the body into the glass cylinder. Partially fill the cylinder with water.

---

* As in using the shellac in the exercise on the lines of magnetic force.

Mark the level of the water by means of a piece of paper about 4 cm. wide and long enough to be wrapped several times around the cylinder. Fasten this marker in position by a rubber band. Be sure that the upper edge of the paper is at the same level all around the cylinder, and that, when looked at horizontally, the line of the paper just touches the lowest point of the meniscus. By means of the thread remove the object, allowing as much of the water as possible to drain back into the cylinder. Run in water with a burette until the level in the cylinder is the same as before. The volume of water run in, as measured on the burette, is the volume of the body.

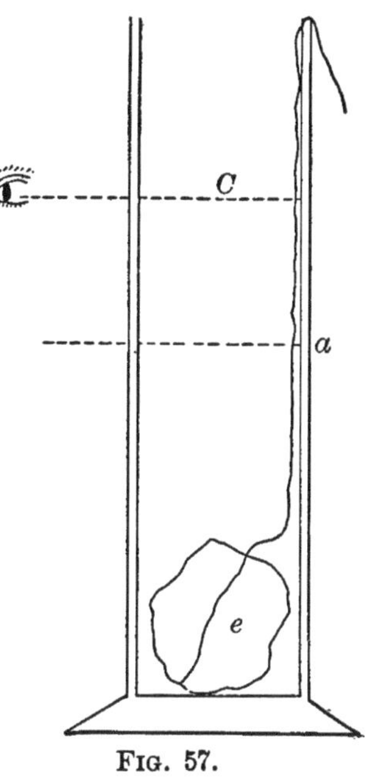

FIG. 57.

Method C. From a burette, run into a dry, empty cylinder any known volume of water,—say 50 cu. cm. —and mark its meniscus by a strip of paper as in *B*. Empty the cylinder, wipe it dry and put in the object whose volume is to be determined. Run in water from the burette until the level is the same as it was before; note the reading of the burette. The difference between the two volumes of water drawn from the burette is the volume of the object. For an object that floats, a sinker must be used. First get the volume of the sinker alone, attach the object to it, and get the volume of both; then by subtracting the volume of the sinker from the volume of the two, the volume of the body alone can be obtained.

Determine by each of these methods the volume of the same body. Compare the results, and state which is the best process in your opinion, and why. As many determinations should be made as time allows, and the results averaged. Arrange as follows :

METHOD A.		1st Trial.	2d Trial.
Vol. water plus body	=	cu. cm.	cu. cm.
" " alone	=	" "	" "
" of body	=	" "	" "
Average vol. by A =		cu. cm.	

METHOD B.	1st Trial.	2d Trial.
Burette-reading after running in,	cu. cm.	cu. cm.
" " before " "	" "	" "
Difference in burette-reading, or volume,	"	
Average vol. by B =	cu. cm.	

METHOD C.	1st Trial : Body out.	2d Trial.
Burette-reading after running in,	cu. cm.	cu. cm.
" " before " "	" "	" "
Volume run in,	" "	"
	Body in.	
Burette-reading after running in,	cu. cm.	cu. cm.
" " before " "	" "	" "
Vol. run in,	" "	" "
Vol. run into empty cyl.,	cu. cm.	cu. cm.
" " " cyl. + body,	" "	" "
" of body,	" "	"
Av. vol. of body =	cu. cm.	

EXERCISE 4.

CROSS-SECTION AND INTERNAL DIAMETER OF A TUBE.

EXPERIMENT.

Apparatus.—Glass tube; burette or equivalent; markers; rubber band; cork; water; meter-stick.

OBJECT.—To find the cross-section and internal diameter of a tube by determining its length and volume.

Manipulation.—*Method A.* Cork one end of the tube, and pour in water enough to cover the cork to a depth of about 1 cm. Carefully mark the level of the water by means of a piece of paper and rubber band, as in the preceding experiment. Run in a known volume of water, say 40 or 50 cu. cm. Place on your tube a second measuring-paper, and move the paper until it just marks the level of the liquid when the tube is held vertically. The distance between the two water-levels, as indicated by the markers, is the height to which the known volume has filled the tube. Pour out the water and place the tube close against the meter-rod, taking care not to disturb the markers. Record the distance corresponding to ab in Fig. 58. Then

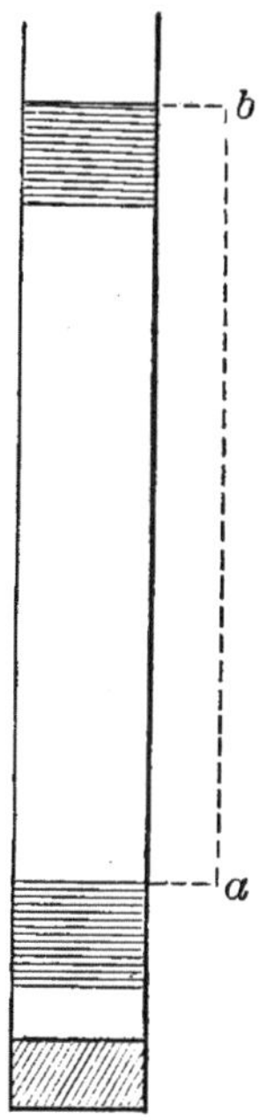

$$\text{Cross-section} = \frac{\text{Volume}}{\text{Height}}.$$

Fig. 58.

Repeat several times with varying volumes. This will reduce the errors due to the uneven bore of the tube. Arrange the results as follows:

No. Trial.	Vol. run in.	Height in Tube.	Cross-section.

Average cross-section =

Method B. Place the two paper markers so that they are a certain distance apart * from upper edge to upper edge, as measured on the scale. Set the tube upright and run in water until the level of the liquid is just at the lower mark. Read the burette. Run in more water to

* The distance between the markers should be an integral number on the scale.

the level of the upper mark, and read the burette again. The difference of the readings is the volume run in between the markers. Repeat several times with the markers at various heights. Arrange results on the plan indicated in the preceding exercise. In Method A, the *volume* being measured to an integral number, the chief liability to error is in the determination of the height. In Method B, the *height* being measured to an integral number, the chief liability to error is in the determination of the volume. Therefore the values obtained by averaging the results of the two methods should be very nearly correct.

To find the diameter, take the average value for the cross-section and substitute it for A in the formula, $A = \pi\frac{D^2}{4}$. Determine the cross-section of a glass tube by both methods, and, if so instructed, compute its diameter, calculating the internal diameter by this formula; carefully measure it according to notes on Measurement, and compare the calculated and the measured results. For comparison, take the average of the diameters of the two ends of the tube. All measurements of length should be made to 0.1 mm., and burette-readings to 0.1 cu. cm. All values of cross-sections to be expressed in sq. cm., and carried out to fourth decimal place. Express the values of diameters in mm. to first decimal place.

DETERMINATION OF WEIGHT.

Introductory.—For exact comparisons of weights, units are required, as in comparisons of lengths or volumes. The ordinary English units are the pound (lb.), and the ounce (oz.), one sixteenth of a pound. The pound is a weight equal to that of a piece of metal in the possession of the government. The metric unit of weight is the gram (g. or grm.), and is the weight of a cubic centimeter of water at a specified temperature. A larger unit, the kilogram (k.), is the weight of 1000 grams. The gram is divided into decigrams (d. or dg.), .1 gram ; centigrams (cg.), .01 gram ; milligrams (mg.), .001 gram. Large weights are expressed in kilograms and decimals of a kilogram ; small weights in grams and decimals of a gram. In a general way, the kilogram, is used in place of the pound in English measure, and the gram in place of the ounce.

The Balance.—This instrument is represented in Fig. 59. *S* is the support, *BB* the beam, and *PP* the pans. Before use the beam should be level, and the bottoms of the pans about half an inch from some surface underneath. The body to be weighed should be placed in the left-hand pan, and

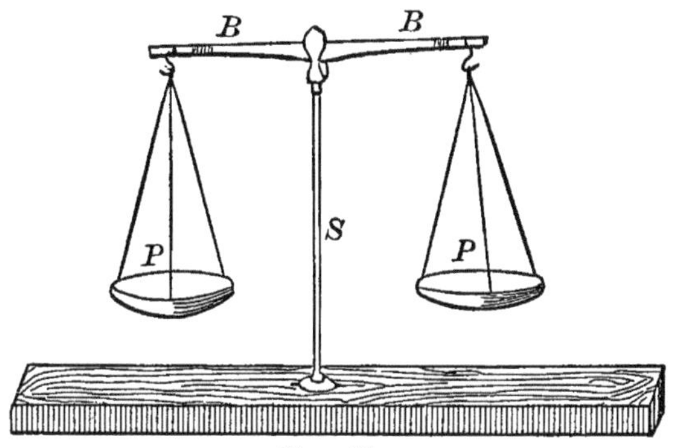

FIG. 59.

weights in the right-hand pan, one by one, until the beam again hangs level. Then the sum of the weights used is the weight of the body. The weights used in laboratories are generally metric. The larger ones down to one gram are made of brass (Fig. 60), and their values are stamped on the top. Weights of less than one gram are made of platinum or aluminium. These have values stamped on them in decimals of a gram.* For example, a weight marked 0.1 would be 0.1 (one tenth) of a gram, or a deci-

* Sometimes the smaller weights are made of wire. The number of parts represented is indicated by bending the wire into a polygon of a corresponding number of sides.

gram; 0.02 means two centigrams, etc. The larger weights are kept in holes bored in a block of wood; the smaller are either in one hole provided with a cover, or in shallow holes covered by a glass plate. In case the smaller weights

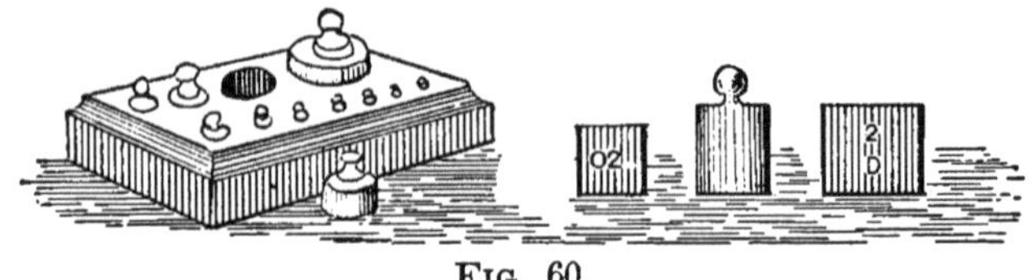

FIG. 60.

are in one hole, they should be taken out and placed upon a piece of paper, marked as follows:

.5	.2	.2	.1
.05	.02	.02	.01

Each weight should be laid so as to cover the mark corresponding to it, and, except when in the scale-pan, should be kept there until the operation is completed. A glance at the uncovered numbers on the card then tells which weights are in the scale pan. Unless very heavy (say above 500 grams), weights should be handled with the pincers provided for this purpose, and not with the fingers. All weights should be placed at once in the proper receptacle when removed from the scale-pan, and never allowed to lie on the table. The large brass weights are handled by the knob on top. Never allow the scales or weights to come in contact with anything that might corrode or injure them.

Weights are of the following numbers and denominations: one five, two twos, and a one. Thus, the weights under one gram would be one-half gram (0.5 grm.), two one-fifth gram (0.2 grm.), and one one-tenth gram (0.1 grm.). Or there are one 100-grm., one 50-grm., two 20-grm., and one 10-grm., etc.

Time is saved by weighing in a methodical manner, and not taking weights at random. Usually the weight of a body can be estimated in round numbers, and the first weight tried should be about that estimated. For example, the body is estimated to weigh about 20 grm. Placing it in the left-hand pan, put a 20-grm. weight in the other. The weight-pan remains in the air, hence 20 grm. is not enough. Add 10 grm. more: this is too much. Replace by 5 grm.: also too much. Replace by 2 grm.: now it is too little. Add 2 grm.: too little still. The weight must be between 24 and 25 grm. Add 0.5 grm.: too little. Add 0.2 grm. more: too much. Replace by 0.1 grm.: still too little. Add 0.05 grm., and the correct weight is found. On adding up the weights in the scale-pan they amount to

	24.00 g
6 dc. or	.60 g.
5 cg. or	.05 g.
	24.65 g.

Weights are usually expressed in grams and decimals of a gram. We would not say that a weight was 24 grams, 6 decigrams, and 5 centigrams, but 24.65 grams; just as we do not say that a thing cost 2 dollars, 4 dimes, and 7 cents, but $2.47.

In getting the weight of a liquid, one method is to first weigh the empty vessel, and then the liquid and vessel together. For example, suppose 10 grams of a liquid are wanted. Weigh a vessel and then add 10 grm. to the weights on the pan and pour the liquid into the vessel until the beam balances. Arrange notes as follows, always placing the weight of the dish alone on the lower line, since it is to be subtracted from the larger weight of the dish and liquid:

Dish and liquid	= 26.249
Dish alone,	= 16.249
Liquid	= 10.000

Another method, called counterpoising, is to place the vessel in one pan, and balance it by some substance, such as sand or shot, in the other. The weight of the liquid alone need not then be determined by actual weighing. It is generally easier to counterpoise a flask or dish than to weigh it accurately.

The Spring-balance.—In this, the weight attached pulls out the spring and registers itself by a pointer on a scale engraved on the brass. The scale is generally graduated up to twenty-five pounds and reads down to one-half pound, every four pounds being numbered. In Fig. 61 the long lines are pound marks, and the shorter lines between them, half-pound marks. In reading, be sure that the eye is directly over the pointer. Always read from the same side of the pointer, which is usually wide enough to cover half a pound on the scale. The scale is quite fine, and a magnifying-glass is often useful. As the pointer is liable to stick, it is well to shake the balance a little before reading. Be careful to hold it so that the rod to which the hook is attached can slide freely. This is especially needful when using the balance to measure force. The balance must be held just in the line of the pull; otherwise it will bind and give incorrect readings. There are some balances which read to 48 pounds. In these the scale only reads to pounds. After being used to reading on a 24-pound balance, one is almost certain to make mistakes in reading on one of the 48-pound balances. Balances reading up to 8 oz. are also used. The scale reads to $\frac{1}{4}$ oz. In using one of these balances, fractions of less than one scale-division should be estimated by the eye. With practice one should be able to read to $\frac{1}{32}$ of an ounce (.037 oz.). Never leave a balance stretched out any longer than necessary, as it injures the spring. If

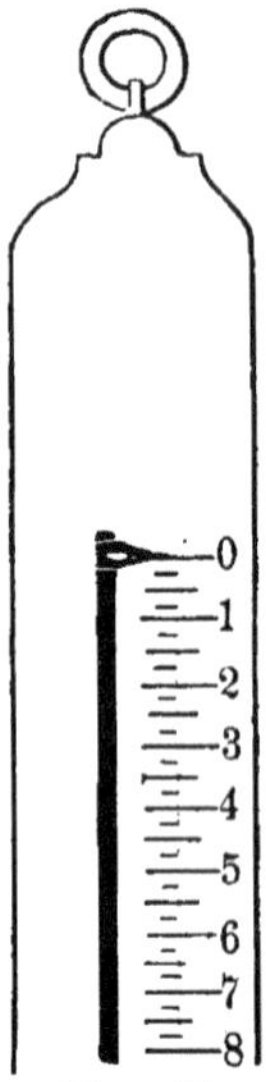

FIG. 61.

the pointer does not stand at 0 when there is no pull on the spring, read the position of the index before beginning to weigh, and subtract (if over 0) from subsequent readings. Thus, if the pointer read one-half pound, all readings are half a pound too high, and half a pound must be subtracted.* A spring-balance is generally used to measure forces; it is then often called a *Dynamometer*, i.e., "force-measurer." When a spring-balance is used for weighing, it should never be held in the hand, but suspended from some solid object.

EXERCISE 5.

PRACTICE IN WEIGHING.

EXPERIMENT.

Apparatus.—Scales; weights; avoirdupois weights or 8-oz. balances (if the latter, one to three students is enough). Several bodies weighing one to two ounces.

OBJECT.—To determine the value of the ounce in grams.

MANIPULATION.—Weigh some body to 0.01 grm. on the scales.† Determine the weight of the same body in ounces by means of a spring-balance or by the scales and English weights. From the following proportion calculate the number of grams to the ounce:

$$\text{wt. in oz.} : \text{wt. 1 oz.} :: \text{wt. in grms.} : x.$$
$$x = \text{no. of grms. to the oz.}$$

Repeat with as many different substances as time allows, average the values so obtained, and tabulate the results as follows:

Body.	Wt. in oz.	Wt. in grm.	Grm. per oz.

* This is often called the *zero error*.

† Select some body weighing 30 to 60 grm. In a general way, the heavier the body, the more accurate the results.

EXERCISE 6.

EXPERIMENT.

Apparatus.—Part I. Meter-stick; blocks of wood; lead-pencils; books; etc.

Part I. Tumbler; measuring-cylinder; water.

Part III. Scales and weights; bodies weighing from 10 g. to 50 g.

OBJECT.—To practise estimating values in the Metric system.

MANIPULATION.—*Length.* Estimate carefully, *by the eye alone,* the dimensions of one of the blocks, the meter-stick being out of sight. Record the estimate. Carefully measure the same distance and record. Try this a number of times, varying the distances measured each time until estimates on several different lengths in succession come quite near to measured values. Record as follows:

Estimated.	Found.	Difference.

The figures in the third column are got by finding the difference between the measured and the estimated distances. If the estimate is less than the true distance, the difference has the minus sign; if greater, the plus sign. It is best to select lengths ending in sharp corners, distances between points marked by a pencil, etc.

Volume. Put some water into the tumbler, estimate the volume in cu. cm. Measure the volume, repeat with various volumes, and record as above. If possible, change the vessel used to hold the water.

Weight. Estimate the weight of a few bodies in grams. Weigh the bodies.

Measure as accurately as possible, but do not try to estimate too closely. It is enough to get within 0.5 cm. of correctness in length, one centimeter in volume, and one gram in weight.

NOTES ON ERRORS.

The results of measurements are never absolutely correct. This is because of imperfections in the apparatus and mistakes of the experimenter. In physical experiments, anything due to these causes, which tends to make results incorrect, is called an *error*. Errors of the experimenter are called *Personal Errors*. For example, in Experiment 1, any mistake in estimating the fraction of a mm., in not reading the scale correctly, or not placing it properly on the table, would be a personal error. Personal errors may be often avoided by using care, and their effect on the result may be still further reduced by averaging. In estimating the fraction of a millimeter, a person is just as liable to over-estimate as to under-estimate ; if he makes a number of determinations, and averages his results, the over-estimates and the under-estimates will tend to neutralize each other, and the average will be nearer the truth. For this reason, in conducting an experiment calling for measurement, the more carefully the work, and the greater the number of determinations, the closer to the truth will be the average result. Errors due to the apparatus are called *apparatus errors*. An error which produces a constant effect, always tending to make the result too high or too low, is called a *constant error*. For example, in Method B, Ex. 3 (Volume of an Irregular Body), when the body was removed from the water it took some water with it. The volume obtained included this water, and was always higher than the truth. Sometimes the value of a constant error can be determined and eliminated. Suppose a spring-balance with no weight attached reads

1 lb., then every reading on the balance will be 1 lb. high, and the true weights can be obtained by subtracting 1 lb. from each weight as given by the balance.

QUESTIONS.—1. Why were a number of determinations made in Ex. 1, and the results averaged? 2. What were some of the personal errors in Ex. 1? What was an apparatus error in Ex. 2? Were there any constant errors in this Exercise? Prepare a list of the personal and apparatus errors in each experiment in this chapter, with suggestions for their avoidance.

EXERCISE 7.

PHYSICAL AND CHEMICAL CHANGE.

Preliminary.—Are physical and chemical changes accompanied by change in weight? As good examples of physical change we may take, first, the solidifying of a liquid (that is, the changing of it from a liquid to a solid when cooled), and, second, the dissolving of a solid, in both cases observing the weight before and after the change. In selecting an example of chemical change, we must take two substances whose union will give a new substance, mix known weights of each, and weigh the products. We can then compare the weight of the new substance formed, with the sum of the weights of the original substance.

EXPERIMENT 1.

Apparatus.—Scales and weights; test-tube and fine wire to suspend it; shavings of wax or paraffine; means of heating; bits of solid caustic potash; water; solutions "No. 1" and "No. 2"; two small vessels in which liquids may be weighed.

OBJECT.—To see if the weight of substances is the same before and after a physical change.

MANIPULATION.—(*a*) Place some scraps of wax or paraffine in a test-tube. Suspend the tube from one end of a balance and gently warm it until the solid melts. Counterpoise and watch while the contents of the tube cool again. What sort of a change have you here? How do the weights of the solid and liquid compare? (*b*) Suspend a test-tube half full of water from one arm of your balance, by means of a thread, and place a small piece of caustic potash on a scrap of paper in the pan. Counterpoise. Drop the solid into the water, leaving the paper on the pan. Watch for any change in weight while the solid dissolves in the liquid. What sort of a change is this? What is your inference? If these experiments illustrate a general law as regards change of weight during a physical change, what do you infer the law to be?

EXPERIMENT 2.

OBJECT.—To see if the weight of substance is the same before and after a chemical change.

MANIPULATION.—Put a small glass vessel on the left-hand scale-pan and record the weight. Add a 10-gram weight to the right-hand scale-pan. Pour some of solution No. 1 slowly into the vessel until it nearly balances (in all probability you cannot get it to exactly balance); change the weight on the scale-pan until you have the exact weight of the vessel + the liquid. Record and find the weight of the solution alone. Weigh in the same manner a part of solution No. 2. The quantity must be exactly determined, but need not be exactly equal to that used of No. 1. Pour the contents of one vessel into the other; note what occurs, and weigh the whole. From this weight subtract the weight of the vessel, and you have the weight of the contents. Compare this weight with the sums of the weights of the two liquids. Have we changed our forms of matter? Is this change chemical or physical? During a chemical change, what is true as regards the weight of matter? This experiment illustrates a general law. From your work, can you state the law? Arrange the results as follows:

Weight of	solution No. 1 + vessel	=	grm.
" "	vessel	=	"
" "	solution alone	=	"
" "	solution No. 2 + vessel	=	
" "	vessel	=	"
" "	solution alone	=	"
" "	two solutions + vessel	=	"
" "	vessel	=	"
" "	two solutions alone	=	"
" "	No. 1	=	
" "	No. 2	=	"
" "	both solutions weighed separately	=	"

DENSITY AND SPECIFIC GRAVITY.

EXERCISE 1.

DENSITY AND ITS DETERMINATION.

Preliminary.—The same weight of matter may take up a great deal of room or only a little. We express this quality of bodies by the words "heavy" and "light." When we say that a body is "light," we mean that the space it occupies, its *volume,* is great compared with its weight, and by the use of the word "heavy" we mean the reverse. In using these words we always refer to some other bodies. Thus, when we say that gas is light, we mean that the relation of its volume to its weight is smaller than that of most bodies. Suppose, now, we compare a stone and a piece of lead. By very rough tests we can determine which is the heavier; but if any exact comparison is desired, the relation of volume and weight must be expressed as a number. To get this number, we must measure the volume and weight of each body, and divide the weight by the volume. The quotient shows how many *times* the volume number is contained in the weight number. A number which, like this, expresses the number of times one quantity is contained in another is called a *ratio.* The ratio of the weight of a body to its volume is called the *density* of the body. By comparing the density number for lead with the density number for stone, we can determine just how many times the lead is *denser* than the stone. Density is sometimes represented by the sign Δ;*

---

* Called *delta.*

and so we can write* $\varDelta = \frac{w}{v}$, or density is the weight of the unit of volume. Would these numbers be the same for different systems of measurement?

EXPERIMENT.

Apparatus.—Scales and weights; 8-oz. balance or English weights; measuring-cylinder; bodies whose densities are to be determined; measuring-sticks (English and French).

OBJECT.—To determine (1) the density of the given substances (*a*) in the Metric system, (*b*) in the English system, ounces to the cubic inch and pounds to the cubic foot; and (2) the order in which the substances should be arranged as regards density.

MANIPULATION.—(*a*) Determine the weight of each body in grams, weighing to 0.1 grm. and its volume in cu. cm. If the body is a geometrical figure, measure the dimensions of the figure and calculate the volume. If the body is irregular, get its volume by one of the methods given in the exercises on mensuration. Find the density by dividing the weight by the volume. Carry this number out to the second decimal place.

(*b*) Determine the weight of each body in ounces by a spring-balance,† expressing fractions of an ounce as decimals. If possible measure the volume in cu. in.; if it is not possible, calculate the volume in cu. in. from the volume in cu. cm., as already found. 1 cu. cm. = 0.061 cu. in.; hence, Vol. in cu. cm. × 0.061 = vol. in cu. in. Calculate the density in English measure, (1) ounces per cu. in., (2) pounds per cu. foot. Arrange the results in two tables, heading the first table, "Table I, French," and the second, "Table II, English, as follows:"

---

* Such an algebraic arrangement of symbols, as a short way of giving a rule, is called a *formula*.

† An 8-oz. balance is the best.

TABLE I, FRENCH.

Body.	Weight.	Volume.	Density.

(2) Arrange the bodies in the order of their densities, that having the greatest density heading the list.

Weigh the liquid in a small vessel in the usual way. In case the vessel is too large for the scale-pan, make a bale of string or wire, like the handle of a pail, and suspend the vessel below the scale-pan, as in Fig. 62. Be careful that the string or wire by which it is suspended does not break, and allow it to drop.

Unit of Density.—We need some standard with which to compare densities. Some density must be taken as the unit, and all densities expressed in terms of that unit. The *density of water* at a specified temperature is used as the unit for solids and liquids, and that of *air* under fixed conditions of temperture and pressure, for gases. In comparing the densities of solids or liquids, we take the numbers obtained by dividing each density by that of water.

Specific Gravity.—The number which shows how many times the density of water is contained in the density of any solid or liquid is called the *Specific Gravity* of that body. Thus, if the density of iron were 437.5 lbs. per ft., its specific gravity would be found by dividing its density, 437.5 lbs. by 62.5 lbs., which is the density of water expressed in the same system. We find that the density of iron is seven times that of water, hence the specific gravity of iron is 7. Calculate the specific gravities of the bodies whose densities you determined in Exercise 1. Are the specific-gravity numbers the same for all the systems of measurement used?

EXERCISE 2.

DETERMINATION OF SPECIFIC GRAVITY.

Preliminary.—In order to determine the specific gravity of a body we could get its density, as in Ex. 1, by dividing its weight by its volume. The density so obtained would be divided by the density of water (if not known, this would have to be found); the quotient would be the specific gravity of the body. Or, again, to simplify the process, we could take *equal* volumes of water and the body under examination. Thus, if any volume of a body weighs 50 grm., and *an equal volume* of water weighs 10 grm., then the density of the body is five times that of the water, and the specific gravity is five.

Suppose we take a bottle, weigh it, then fill it completely full of water and weigh it again. The weight of the bottle subtracted from the combined weight of the bottle and water gives the weight of the water. Suppose, now, we empty the water completely out, and fill the bottle with the liquid whose specific gravity we wish to determine. Then, since we have the same bottle, and have filled it completely full, we have the same volume of liquid that we took of water. If we weigh the bottle and liquid together, and subtract the weight of the bottle, we get the weight of the liquid. Having the weights of equal bulks of water and the liquid, we can get the specific gravity of the liquid by dividing the weight of the liquid by the weight of the water.

EXPERIMENT.

Apparatus.—Scales; weights; specific-gravity bottle; liquid whose specific gravity is to be determined; water.

OBJECT.—To determine the specific gravity of a liquid by the method of the "Specific-gravity Bottle."

MANIPULATION.—Weigh the bottle with the stopper in. Be sure that it is perfectly dry. Fill it to the top with the liquid, and, holding it over some vessel to catch the overflow, slowly insert the stopper squarely, as in Fig. 61. During this operation the liquid should run over the rim of the bottle all around. With care, the bottle will be completely filled. To test this, turn the bottle upside down and see if any air bubbles appear. If they do, the bottle is not entirely full, and the operation of filling must be repeated. Wipe the outside of the bottle dry and weigh it. Pour the liquid back into the "stock-bottle," rinse the gravity-bottle, fill it completely with water with the same precautions as before, and weigh it. Arrange results as follows:

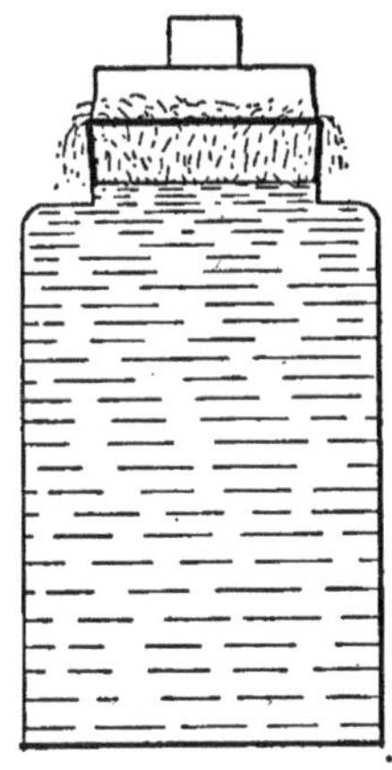

FIG. 61.

Wt. bottle + liquid =	Wt. bottle + water =
" " alone =	" " alone =
" liquid " =	" " " =

$$\text{Sp. grav.} = \frac{\text{Wt. of liquid}}{\text{Wt. of water}} =$$

QUESTIONS.—1. What is the principle of this experiment? 2. Why must you be sure that the bottle is completely full each time? 3. What do you consider the most important error that is likely to be made? 4. Why must the same bottle be used each time? 5. Why should the bottle be rinsed out before filling it with water?

EXERCISE 3.

THE WEIGHT LOST BY A BODY WHEN IMMERSED IN A LIQUID.

Preliminary.—We know that a body does not weigh as much under water as it does above, and that when it is completely immersed it displaces a volume of water equal to its own volume.* The object of the following exercise is to see if there is any connection between this loss of weight and the weight of the water displaced. On what principle must we work in order to be able to make this comparison?

In order to get the loss of weight, we can weigh the body in air and then in water. The difference is the body's loss of weight. To weigh the water displaced, we can fill the specific-gravity bottle completely full of water, and weigh it and the body together. Then if we put the body into the bottle it will crowd out a volume of water *equal to its own volume.* If, after inserting the stopper, we weigh the bottle with the remaining water and the body, the weight will be less than before by the weight of the water crowded out. This difference is the *weight* of the displaced water. Having found the body's loss of weight and the weight of the displaced water, we can see if there is any connection between the two facts.

EXPERIMENT.

Apparatus.—Specific-gravity bottle; water; scales and weights; solid body; fine wire for suspension; tumbler.

OBJECT.—To compare the weight lost by a body when immersed in a liquid, with the weight of a volume of the liquid equal to the volume of the body.

MANIPULATION.—Fill the specific-gravity bottle full of water and place it in the scale-pan. Place the body in the same pan and weigh both. Take the stopper out of the bottle, put in the body, replace the stopper, wipe the bottle dry and weigh again. The difference in the weights is the weight of the water crowded out of the bottle when the body was put in, and hence the weight of a volume of water equal to the volume of the body.

* Provided, of course, the liquid does affect on the solid.

To get the body's loss of weight in a liquid, suspend it by a thread, or, better, a fine wire, which, as shown in Fig. 62, passes through holes in the scale-pan and box, and is attached to the scale arm-hook. Weigh. Put a tumbler of water under the box and adjust it so that the body is

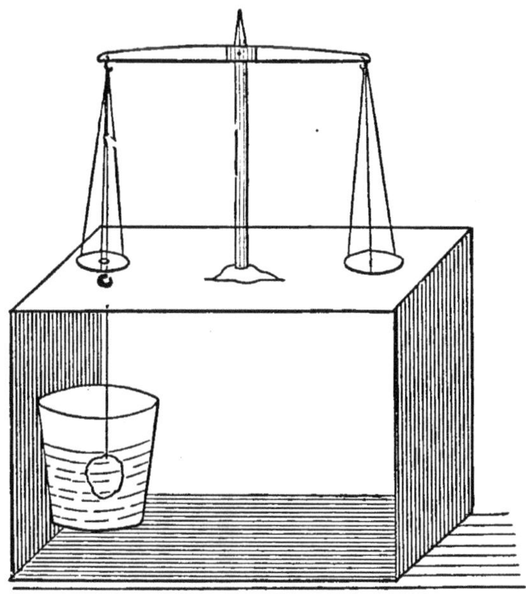

Fig. 62.

completely immersed in the liquid and touches neither the sides nor the bottom. Weigh the body, being careful to keep it entirely under water. If time allows, repeat with some other liquid. Record results as follows :

Bottle + water + body outside	=
" + " + " inside	=
Weight of water displaced	=
Weight of body in air	=
" " " " water	=
	———
Loss of weight	=

EXERCISE 4.

THE DETERMINATION OF SPECIFIC GRAVITY BY IMMERSION.

Preliminary.—It is evident that the method of the specific-gravity bottle cannot be used for solids, and so the principle of the preceding exercise is employed in determining the weight of the water equal in bulk to the solid whose specific gravity is required. By weighing the solid in air and then in water we can get its loss of weight, which last, we have found, is also the weight of an equal volume of water. We then have the weight of the body and the weight of an equal volume of water, and can calculate the specific gravity.

The same principle can be used in getting the specific gravity of liquids. By ascertaining the loss of weight of a solid in water, we get the weight of a volume of water equal to the volume of the solid. By finding the loss of weight of the same solid in the liquid whose specific gravity is to be determined, we get the weight of a volume of the liquid equal to the volume of the solid. The body being the same, we have the weights of equal volumes of the liquids, and can compute the required specific gravity.

EXPERIMENT 1.

Apparatus.—Scales and weights; body and wire for suspension; tumbler; liquid whose specific gravity is to be determined; water.

OBJECT.—To determine the specific gravity of a solid not affected by water, and of a liquid, by the method of "Double Weighing."

MANIPULATION.—*Part I.* Suspend the solid by a thread or fine wire, as in the preceding exercise (Fig. 62). Care must be taken not to let the solid touch the glass at any point.* Weigh the solid as closely as can be done on the

* With an irregular body, a glass stopper for instance, some judgment must be exercised regarding the manner in which the body is suspended.

scales. Record the weight. Fill the tumbler with water and weigh again, taking care that the body is entirely immersed and touches neither the sides nor the bottom of the glass, and that the suspending thread does not touch the sides of the hole in the box. Record as follows:

Wt. in air =
" " water =

Loss of " " " =

$$\text{Sp. gr.} = \frac{\text{Wt. in air}}{\text{Loss of wt. in water}} = \text{[Carry out to 2d dec. pl.]}$$

Part II. After weighing the body first in air and then in water, and recording the weights, empty the water out from the tumbler, fill the tumbler with the given liquid, and weigh the body again, using in all cases the same precautions as in Part I. Record results as follows:

Wt. in air =	Wt. in air =
" " water =	" " liquid =
Loss of " " " =	Loss of " " "

$$\text{Sp. grav.} = \frac{\text{Loss wt. in liquid}}{\text{Loss wt. in water}} =$$

How does this experiment compare in principle with the method of the specific-gravity bottle?

If Part II is performed immediately after Part I., the solid's loss of weight in water is already known, and we have only to weigh the solid in the liquid whose specific gravity is required. The record would then take the following form:

[From Part I.]	Wt. in air =
	" " liquid =
	Loss of " " " =
[From Part I.]	" " " " water =

$$\text{Sp. grav.} = \frac{\text{Loss of wt. in liquid}}{\text{Loss of wt. in water}} =$$

EXPERIMENT 2.

Apparatus.—Scales and weights; water; tumbler; body lighter than water; sinker; thread or fine wire.

OBJECT.—To determine the specific gravity of a body that will float in water.

MANIPULATION.—Weigh the body in air. Attach the sinker close to the body with thread or wire; suspend the two, as in the previous experiment, and with the same precautions determine the weight of the two in air. Determine the weight of the two when immersed in water, in the same way as in the previous experiment. Remove the body and determine the weight of the sinker alone when immersed in water.* Arrange results as follows:

Sinker + body in air =
" alone in air =
Body in air =
Sinker in water =

CALCULATION.—Arrange calculations as follows:

Sinker + body in air =	Sinker in air =
" " " water =	" " water =
Loss of wt. of both in water =	Loss of wt. of sinker =

Loss of wt. of body + sinker =
" " " " sinker alone =
" " " " body " =

$$\text{Sp. grav.} = \frac{\text{Wt. of body in air}}{\text{Loss of wt. of body in water}} =$$

* The number of operations involved in this experiment may be reduced by taking the data obtained in Experiment 1, which may be done by using for a sinker the body whose specific gravity was there determined. It is then only necessary to ascertain the weight of the sinker plus the body in air and the loss of weight of both in water.

Questions.—1. In these experiments, is the weight of the equal volume of water actually measured at all? 2. Why are you justified in taking the loss of weight as the weight of a bulk of water whose volume equals that of the body? 3. State the principles of this experiment. 4. How does it differ from that used in getting the specific gravity of a liquid?

EXERCISE 5.

LIQUID PRESSURE DUE TO WEIGHT.

Preliminary.—For investigating the conditions affecting liquid pressure, we use the apparatus shown in Fig. 63. A pressure-gauge is made of a glass funnel, *a,* whose end is

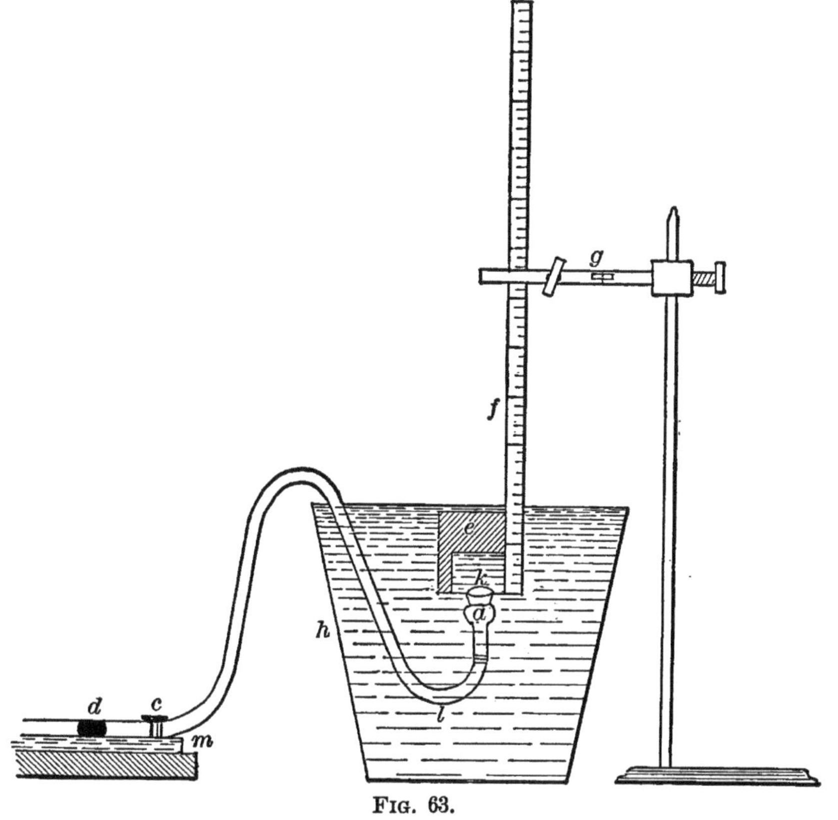

FIG. 63.

covered with thin rubber, *k.* From the other end, a rubber tube, *l,* connects the funnel with the glass tube, *cd,* which is attached to the scale, *m*, and contains a drop of water, *d*, to act as an index. Changes of pressure will cause the rubber, *k*, to bulge more or less, and this will cause a motion of the index which can be read upon the scale. This gauge is hung from the block, *e*, by a wire on which it turns, so that it may be made to face in any direction and the centre of *k* remain at the same depth. To this block is attached a meter-stick, *f*, which indicates the depth. The apparatus is held by a clamp, *g*, and is placed in a pail of water, *h*. The scale *m* lies flat upon the table.

EXPERIMENT.

Apparatus.—As shown in Fig. 63. Water and pail (or equivalent); measuring-cylinder (or something that can be used to produce changes in depth directly over the gauge-face.)

OBJECT.—To investigate the conditions affecting the pressure of a liquid upon a surface immersed in it.

MANIPULATION.—Before commencing the experiment it is necessary to know how the movements of the index correspond with changes in pressure. The gauge being in the water, press on its face gently with the finger and note the direction of the movement of the index. Remove the pressure and note again. Record the direction of movement for increased and decreased pressure, as follows:

Increased pressure-index moves
Decreased " " "

Part I. Effect of depth. The gauge-face being 1 or 2 cm. below the level of the liquid, note the position of one end of the index, and the depth of the gauge. Increase the depth 4 or 5 cm. (most easily done by setting the clamp lower down on the rod), read the depth and the position of the index. Be careful to always read from the same end of the index, to clamp the apparatus firmly at each depth, and to wait for the liquid to come to rest before reading. In this way take readings at various depths, working first down to as near the bottom of the vessel as is possible, and then up again to near the surface. Tabulate results as follows:

Depth.	Index-reading.

From the study of these figuures, place in the note-book an inference regarding the effect of depth on pressure.

Part II. Effect of direction. Adjust the apparatus so that the gauge-face is 6 or 7 cm. below the surface. Putting the hand into the water, take the rubber tube *gently* between the thumb and finger at a point just below where it joins the glass, and slowly turn the gauge-face in various directions, reading the index for each direction. Take care that there is plenty of slack rubber tube, and that it does not "kink" anywhere. Make three or four trials at this depth, and then repeat at some other. Record results as follows:

Depth.	Direction of gauge-face.	Index-reading.

What inference?

Part III. Effect of distance from sides of vessel at same depth. Gently set the gauge at various points at the same level by moving either the arm of the clamp horizontally or the pail under the apparatus. In each case wait for the liquid to come to rest, read, and record as follows:

Depth.	Position relative to vessel sides.	Index-reading.

What inference?

Part IV. Effect of depth of liquid directly over the immersed body. Tip the apparatus sideways as shown in Fig. 64, and clamp it there. Bring the bottom of a glass

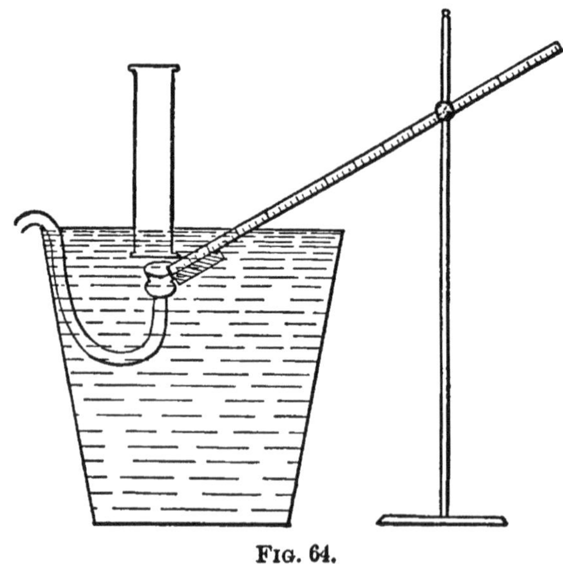

FIG. 64.

cylinder down over the gauge-face, as in the figure. This reduces very much the depth and amount of the liquid directly over the gauge-face, but does not change the general level of the liquid to a noticeable degree.* Immerse the bottom of the cylinder to various distances above the gauge-face. Read the index each time. At each point hold the cylinder steady and wait until the water in the pail has come to rest before reading the index. To increase the depth of liquid directly over the gauge-face, submerge the cylinder completely and invert it. Holding the inverted cylinder by the bottom, raise it until the lower end is only one or two cm. below the level of the liquid. So long as no air enters, the water will remain in the cylinder, and by bringing it over the gauge-face, as in Part IV, the depth of water there may be made considerably greater

* Of course, the larger the pail, the less this error amounts to.

than elsewhere in the pail. The admission of air into the cylinder, by raising one edge for an instant above the level of the liquid, will enable you to change the depth of water in it. In this way try various depths. Record results as follows:

Depth of Gauge-face below General Surface.	Depth Liquid directly above Gauge-face.	Index-reading.

What inference?

Write out a summary of what you have learned about all the conditions affecting the pressure that the weight of a liquid causes it to exert on an immersed surface.

EXERCISE 6.

SPECIFIC GRAVITY OF LIQUIDS BY THE METHOD OF BALANCING.

Preliminary.—In the following exercise we wish to determine the specific gravity of a liquid by the method of balancing. This method is based on the laws of liquid pressure. Suppose we have two tubes, connected as at *a* in Fig. 65. Pour some liquid into one tube, and a liquid that will not mix with it into the other. Then, when the liquids have come to rest, we shall have two columns balancing each other. Where the two liquids join, *a*, there are two pressures—a downward pressure due to the weight of the liquid above, and an upward pressure due to the liquid in the other tube. When the

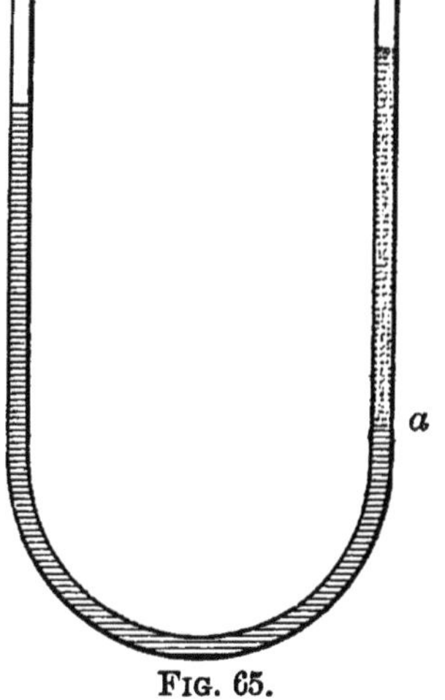

FIG. 65.

liquids have come to rest, we know that these pressures are equal. Call the upward pressure P, and the downward pressure P'; then $P = P'$. But we know that

$$P = a \times D \times \Delta,$$

and

$$P' = a' \times D' \times \Delta',$$

calling a', D', and Δ' the values for the second liquid. So

$$a \times D \times \Delta = a' \times D' \times \Delta'.$$

But as the pressures come together in the same tube at the same point, $a = a'$, so when two liquids come to rest in communicating vessels,

$$D \times \Delta = D' \times \Delta',$$

or, the depth times the density of one equals the depth times the density of the other. Hence, to determine the required density, we have only to put into one tube a liquid of known density, into the other the liquid under examination, measure the depths of the two liquids, and divide the product of the density and depth of the former by the depth of the latter, or $\Delta' = \frac{D \times \Delta}{D'}$.

EXPERIMENT.

Apparatus.—As shown in Fig. 66. Funnel; water and oil—with vessels to contain them; meter-stick; solution of copper sulphate or equivalent.

OBJECT.—To determine the specific gravity of a liquid by the method of balancing.*

---

* Observe that this special method can only be used for liquids that do not mix.

MANIPULATION.—By the aid of the funnel, pour water into the tube until it stands about half-way up in both branches. Make sure that no air remains in the rubber tube by drawing the tube through the fingers pressed together. Slowly pour the oil into the larger tube, at first letting it run down the sides so as to accumulate quietly on top of the water, and continuing until the column is 50 cm. or so long. Care must be taken that the junction of the liquids does not get below the glass into the rubber tube. In case this happens, it can sometimes be remedied by pouring more water into the water tube. Avoid getting the oil into the water tube.

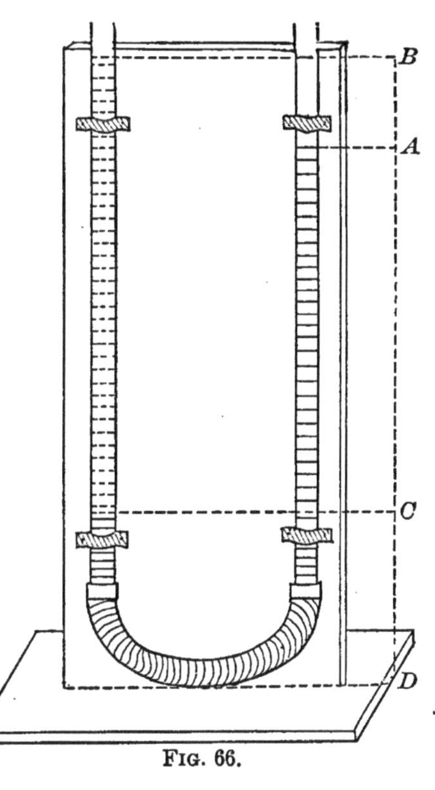

FIG. 66.

When the liquids have come to rest, starting from some horizontal line below the tubes, as D, Fig. 66 (the floor, the base-board, or the table will do), measure the distance:

(*a*) from the reference-line to the top of the water column, or *DA*;

(*b*) from the reference-line to the top of the other column, *DB*;

(*c*) from the reference-line to the junction of the liquids, marked *DC*.

CALCULATION. $BD - CD =$ Height of the column of the liquid whose specific gravity is to be determined, and

$AD - CD =$ Height of the water column producing the same pressure on the same area. Therefore

$$\text{Sp. grav.} = \frac{A - C}{B - C}.$$

Make three or four determinations with various heights, changing the heights by pouring in either water or oil. Tabulate results as follows:

TABLE I.

BD	*AD*	*CD*	*BD* − *CD*	*AD* − *CD*	Sp. Grav.

If time allows, repeat, using solution of copper sulphate as the liquid whose specific gravity is to be determined, and the oil as the liquid of known density, taking for its specific gravity the average of three trials with water.

EXERCISE 7.

EXPERIMENT 1.

Apparatus.—Loaded test-tube with cork and weights; measuring-cylinder; water; other liquids of known density.

OBJECT.—To compare the weight of a body that will float with the weight of the liquid displaced by the body when floating.

MANIPULATION.—Weigh the test-tube; pour about 30 cu. cm. of some liquid into the measuring-cylinder, carefully read the volume, and holding the test-tube by the cork, let it slip gently into the liquid until it floats. Be careful that the measuring-cylinder is dry above the level of the liquid after the test-tube is inserted. In order that the body may float freely, tap the measuring-cylinder several times with a lead-pencil. Again read the volume of the liquid. The difference in the readings gives the volume of liquid displaced by the body, and this volume multiplied by its density gives the weight. Repeat the determination with two other liquids. Record results as follows:

Vol. before.	Vol. after.	Vol. displaced.	Weight displaced.	Liq. used.	Wt. of Body.

EXPERIMENT 2.

Object.—To determine the specific gravity of a liquid by the use of a floating body.

Manipulation.—With the apparatus above, determine the specific gravity of another liquid, writing out in the note-book a complete statement of the special method, etc.

Questions.—1. State the principle of these experiments. 2. How does it differ from that on which the preceding exercises were based? 3. How does the loss of weight of the body compare with its weight in air? 4. Do you think this fact had any connection with the floating of the body?

EXERCISE 8.

ATMOSPHERIC PRESSURE AND THE BAROMETER.

EXPERIMENT.

Apparatus.—Barometer tube; clamp; handkerchief or cloth; mercury (about 1 k.) funnel; feather on wire for removing air-bubbles; dish to hold mercury; meter-stick; scales and weights; beaker glass for weighing mercury.

Object.—To measure the pressure of the atmosphere.

Manipulation.—Clamp the barometer-tube in an upright position, closed end down, placing under it some soft substance, as a handkerchief or towel. By the aid of the funnel fill the tube about half full of mercury, then gently insert the feather, and, by turning the wire, remove the bubbles of air which adhere to the side. Add 10 or 15 cm. more of mercury and repeat the removal of the air-bubbles. Proceed in this way until the tube is filled with mercury and contains no air. Have ready the dish containing the mercury filled to a depth of 3 or 4 cm.; grasp the tube near the top with the right hand, placing the thumb firmly over the opening. With the left hand unclamp the tube, grasp it near the bottom, and (keeping the thumb firmly on the open end) inverting the tube, place the end in the vessel below the level of the mercury. Remove the thumb and again clamp the tube vertically, being sure that the clamp takes the weight of the tube, which must not rest on the bottom of the vessel. Observe and record what happens.

When the mercury column has come to rest, carefully measure its height above the level of the mercury in the vessel. Placing the thumb loosely over the lower end of the tube and holding the tube as before, raise it gently until its lower edge is just below the level of the mercury, then press the thumb firmly on the bottom, lift the tube out, incline it in a nearly horizontal position, and allow the mercury to run very gently into the thin glass vessel by admitting air, a few bubbles at a time, into the bottom of the tube. Be sure that none of the mercury is spilled. Weigh the vessel with the mercury, and also weigh the empty vessel and compute the weight of the mercury. This gives the pressure of the atmosphere on each area equal to that of the cross-section of the tube.

CALCULATION.—To find the cross-section of the tube, divide the weight of the mercury by 13.6 to get the volume, and this quotient in turn by the height, as in the experiment on the Cross-section of a Tube. To find the pressure per sq. cm., make the proportion A : 1 sq. cm. ::

$W : W'$. Where A is the area of the tube, W is the weight of the mercury, and W' is the weight per sq. cm.

EXERCISE 9.

SPECIFIC GRAVITY OF TWO LIQUIDS BY BALANCING AGAINST THE ATMOSPHERIC PRESSURE.

Preliminary.—If we dip the lower end of a tube into a liquid and remove a portion of the air from the tube, the tension of the air remaining becomes less than that of the air outside, and a column of the liquid is forced up the tube, until its weight, plus the tension of the air above it, produces a pressure equal to the tension of the air outside. If we place the tube in another liquid and withdraw the same amount of air, then the second liquid-column formed has to furnish the same pressure as the first. The two columns, instead of being balanced against each other, as in Exercise 6, are balanced against the same pressure, that of the atmosphere, and must be of equal weight. So, as before,

$$\Delta \times D = \Delta' \times D'.$$

In the following exercise we wish to determine the specific gravity of a liquid by this method. In order to have the same reduction of tension in each tube, both are connected with the same vessel, as in Fig. 67, and the air drawn out of that. If one liquid has a known density, we can determine the specific gravity of the other, by the same calculation as in Exercise 6.

EXPERIMENT.

Apparatus.—**Form shown in Fig. 67. Vaseline; meter-stick; water; solution of copper sulphate and some other solution.**

Object.—To determine the specific gravity of a liquid that will mix with water, by the method of balancing.

MANIPULATION.—Arrange apparatus as in Fig. 67. Place water in one tumbler and in the other the copper sulphate. By applying the lips to the glass mouthpiece *a*, suck some air out of the bottle *B*, thus causing a column of liquid to rise in each tube. Without removing the mouthpiece from the lips, compress the tube tightly with the left hand, and while holding it in this manner, replace the mouthpiece by the solid plug (which may be covered with vaseline). On releasing the tube, the columns of the liquids will stand at a certain height in each. By bringing a scale alongside the tube, the height of each column from the level of the liquid may be obtained. The specific gravity is equal to the height of the water column divided by the height of the other column. Repeat several times with different heights. Tabulate result:

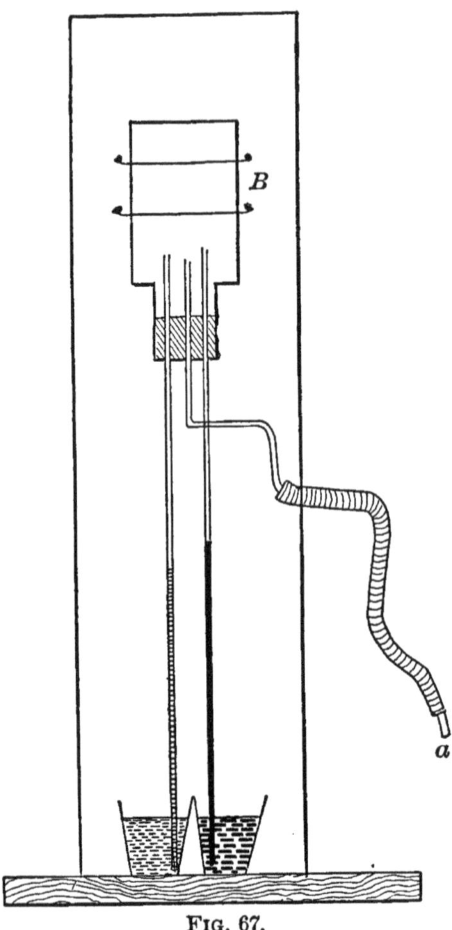

FIG. 67.

Height of Water.	Height of other Liquid.	Specific Gravity.

Replace the water by some other liquid, and, taking the

copper sulphate as the liquid of known density, determine the specific gravity of the other liquid.

QUESTIONS.—How does this method differ in principle from Exercise 6? With sufficiently long tubes, could determinations of specific gravity be made in this way by entirely removing the air? Suggest a method of determining specific gravity based upon Exercise 8.

The Bunsen Burner.—The instrument commonly used for heating bodies is a burner, called the *Bunsen Burner*. This instrument (Fig. 68) is connected with the gas-pipe by a rubber tube, and consists of a pipe, *a*, provided with holes near the bottom, which can be closed, if desired, by turning the sleeve, *S*. The gas enters the bottom of the tube through a small opening, and, when lighted at the top of the tube, burns with a very hot blue flame, free from smoke. On shutting the air-holes, the flame becomes yellow and smoky. Never use the yellow flame except when specially instructed to do it. It will cover with a coating of lamp-black whatever is put in it. When turned down low, the flame sometimes runs down to the bottom of the tube, and burns at the point where the gas enters.* This is called "backing-down," and is often indicated by the flame taking a green color. Backing-down should be stopped at once, as it will make the lamp very hot, sometimes even hot enough to melt the rubber tube. A smart blow of the fist on the rubber tube as it lies on the table will often cause the flame to jump to the top of the tube. If this fails, the gas must be turned off and relighted.

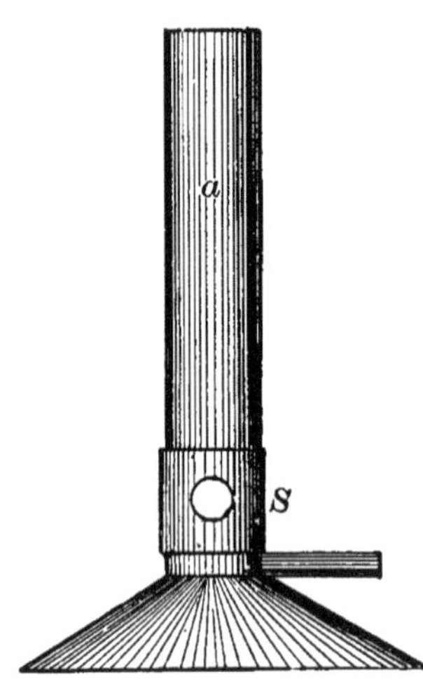

Fig. 68.

* Partially closing the sleeve when the gas is turned low tends to prevent this.

Precautions in Heating.—If a flask is nearly full of water, it is fairly safe to heat it directly, but care must be used that the flame strikes the glass nowhere above the level of the water. Test tubes are usually heated directly in the flame, with the same precautions. They may be held by a strip of paper, doubled three or four times, and passed around the tube near the top.* The bottom of any vessel to be heated should usually be held about three inches above the top of the burner. The safest way of heating a glass flask is by means of a shallow saucer of sheet-iron, filled with sand, in which the bottom of the flask rests. Such an arrangement is called a *sand-bath*, and, when the burner is placed under it, should be supported three or four inches above the top of the tube. Instead of a sand-bath, a piece of wire-gauze is often used. This is placed upon the ring of the ring-stand, and the bottom of the flask allowed to rest upon it. *Never place your hand where, should the vessel heating break, the hot contents can fall upon it.*

EXERCISE 1.

HOW HEAT TRAVELS.

Preliminary.—In the following exercise we wish to observe what happens when one portion of a body is raised to

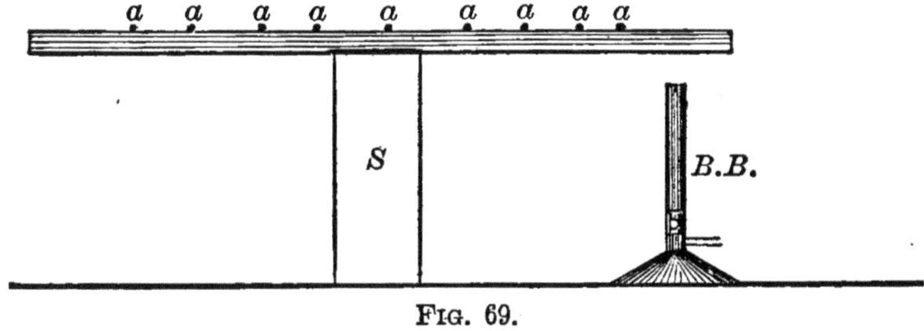

FIG. 69.

a higher temperature than the other portions. We may use the apparatus shown in Fig. 69, which consists of a

* Essentially as in Fig. 48.

rod held by the support *S*, so that one end may be heated by the burner *BB*. Make some pellets of wax a little larger than the head of a pin, and place them at regular intervals along the upper side of the rod, as *aa*. Any considerable rise in temperature at any point on the rod will be indicated by the melting of the pellet there. We must try several rods of different materials, and vary the conditions by bringing two different substances in contact, instead of using different parts of the same substance. Finally, we must try a liquid. The case of a liquid is a little different, because its particles are free to move among themselves, and we shall require some means of observing such motion if it occurs.

EXPERIMENT 1.

Apparatus.—**Exp. 1-3. Rod to be heated, with support; wax or paraffine; means of heating the rod. Exp. 4. Incandescent lamp that can be lighted. Exp. 5. Rods of wood, glass, and iron, with means of heating them. Exp. 6. Rod of Exp. 1; thin glass vessel; water; means of heating. Exp. 7. Ring-stand; lamp; wire-gauze or sand-bath; water; some crystals of potassium permanganate.**

OBJECT.—To observe what happens when one portion of a solid is kept at a higher temperature than the rest.

MANIPULATION.—The rod being fastened so that the end to be heated is about 2 cm. above the top of the Bunsen burner, close the air-holes of the burner, light it, and turn off the gas until the flame is about .5 cm. high.* Heat one end of the rod with this low flame, and record your observations. If possible, note the time at which the melting of each piece of wax begins. Write out all you have learned regarding what takes place when a portion of the rod is kept at a higher temperature than the rest of the rod.

EXPERIMENT 2.

OBJECT.—To study the distribution of the heat in the rod.

---

* This gives the yellow, smoky flame.

MANIPULATION.—Remove the burner, and by the fingers or any convenient method test the temperature at different points of the rod, including the ends. Draw a line to represent the rod, and illustrate the distribution of the heat by a line drawn around it.

EXPERIMENT 3.

OBJECT.—To find if the rod loses heat.

MANIPULATION.—Replace the burner, open the air-holes, turn on the gas, and heat one end of the rod about two minutes. Remove the lamp and bring your hand near, but not touching the end that was heated. Does the rod seem to be losing heat? Is this loss in every direction? To test this latter point, hold the hand about .5 cm. from the heated end of the rod, above, below, on either side, and horizontally from the end. Write out what you have learned regarding the loss of heat.

EXPERIMENT 4.

OBJECT.—To observe if the results in Exp. 3 can still be obtained in the absence of air.

MANIPULATION.—Observe an Edison lamp ; turn on the current and see if the heat from the hot carbon reaches your hand through the space in the globe from which the air has been exhausted. Record your answer to the question.

EXPERIMENT 5.

OBJECT.—To find if all bodies behave in a similar manner when one portion of them is heated.

MANIPULATION.—Repeat Exp. 1 with rods of wood, iron, glass, etc., recording carefully the results in each case.* See if you can class together any substances that behave alike in this respect. Record results in tabular form.

* Or the experiment can be tried by holding one end of the rod in the hand, heating the other end in the flame, and observing whether the end in the hand becomes heated or not.

EXPERIMENT 6.

OBJECT.—To observe what happens when a heated body is brought in contact with a cooler body.

MANIPULATION.—Place about 50 cu.cm. of water in a glass vessel and note the degree of warmth by means of the finger. Heat one end of the copper rod for a few moments, plunge it into the water, stir it around, withdraw it, and again test warmth of the water. State your inference.

EXPERIMENT 7.

OBJECT.—To observe what happens when one part of a liquid is heated hotter than the rest.

MANIPULATION.—Support the thin glass vessel used in Exp. 6 on a ring-stand, by means of a wire-gauze; fill it three-quarters full of clean water. Have ready the burner turned down low, and adjust the ring-stand so that the vessel is supported about 3 cm. above the top of the burner. Drop into the water four crystals of the pink substance given you (which in dissolving colors the water), and at the same instant place the lamp under the vessel. Any motion of the liquid will be indicated by the motion of its colored portions. Watch carefully and draw a diagram representing your observations. Compare as carefully as possible the behavior of a solid with that of a liquid when one part is heated more than another.

Definitions.—The process by which heat is transferred from one part of a body to another part, or from one body to another in contact with it, is called *Conduction*. The process by which a body loses heat when in contact with no other body, as in the case of the electric light, is called *Radiation*. The process by which heat is distributed through a liquid or gas, as in Exp. 7, is called *Convection*. The condition of a body as regards its ability to give up heat is called its *Temperature*. The body giving up heat is said to have the higher temperature. The word "temperature" is used to indicate the relative degree to which a body possesses the property of causing the sensation that we call heat. It is always used with reference to the condition of some other body taken as a point of comparison.

EXERCISE 2.

TESTING THERMOMETERS.

Preliminary.—The fact that bodies expand when their temperature is raised and contract when it is lowered, is made use of in constructing instruments to measure changes in temperature. These instruments are called *thermometers.* They usually consist of some substance so arranged that changes in its volume may be observed on a scale. The instrument generally used in laboratory work is called a *chemical thermometer.* The substance used is mercury. It is contained in a glass bulb connected with a fine tube with thick walls. The scale is engraved on the walls, every ten degrees being numbered. The centigrade scale is generally used. At the top of the instrument is a small glass eye by which it may be suspended. Such thermometers are usually provided with a case in which they should always be kept when not in actual use. The glass of the bulb is very thin, and great care should be used not to break it. When a thermometer is used to determine the temperature of a liquid, the bulb should not be allowed to come in contact with the sides or bottom of the containing vessel. When a thermometer has been at one temperature and is to be exposed to a very different one, as in changing from ice-water to steam, hold it in the air a moment before exposing it to the new temperature. For changes of a few degrees this precaution is not necessary. In reading a chemical thermometer, a white card held behind the glass makes the position of the top of the mercury column much more distinct.

After thermometers have been used for a time, the position of the mercury at the temperature of melting ice does not always agree with the 0 point on the scale, and sometimes the position at 100° also changes. Hence the thermometer used in the following exercises should be tested, in order that, if necessary, corrections may be made in its readings. The correctness of the 0 point is tested by placing the thermometer in *melting* ice (whose temperature is always 0); the 100° point is tested by immersing the thermometer in steam (whose temperature is known).*

EXPERIMENT.

Apparatus.—Thermometer to be tested; ice or snow; water; flask and ring-stand, and wire-gauze or sand-bath; tumbler; clamp; delivery-tube; large glass tube 3 or 4 cm. longer than thermometer, corked at one end, the cork having a hole for the thermometer and another holding a piece of small glass tube for connecting with boiler.

OBJECT.—To test the correctness of the points on the scale of the thermometer which correspond to the temperatures of melting ice and steam.

MANIPULATION.—*Part I.* Place the thermometer in the centre of some pounded ice or snow in a tumbler, the zero-point† on the scale being just exposed. Allow it to remain there until the mercury has ceased to fall. Record the point on the scale at which the mercury comes to rest. This point is the true zero-point on the scale.

Part II. Clamp the large tube in a vertical position, corked end up. Connect the delivery-tube from a flask about two-thirds full of water with the small tube in the cork. Thrust the thermometer through the other hole in the cork until the 100° point is just above the cork. Boil the water in the flask, thus surrounding the thermometer

* Of course thermometers do not need testing every year or with every class.

† These instructions assume that the thermometer has a centigrade scale.

with steam, and note the reading of the mercury column when it comes to rest. Draw the thermometer up until the bulb only is below the cork, and see if it makes any difference in the reading whether all the mercury is heated or not. Read the barometer. The temperature of the steam corresponding to the atmospheric pressure may be calculated by calling the temperature corresponding to 760 mm. 100°, and adding 1° for each 27 mm. above 760, or subtracting 1° for each 27 mm. below. Record the thermometer-reading, the true temperature of the steam, and the difference.

EXERCISE 3.

TEMPERATURE AND PHYSICAL FORM.

Preliminary.—We already know that when the temperature of a solid is raised sufficiently, the solid changes to a liquid, and at a still higher temperature to a gas. That is, the physical form of a body is affected by its temperature. In the following exercise we wish to find out all we can about what goes on when a body is heated. Let us heat a body and note its changes in temperature, and also watch for any other changes. Ice is a good substance to work with. If we put some ice in a vessel, heat it, and note the temperature at regular intervals, we can see if there is any definite connection between temperature and physical form. At the beginning, some water must be added to the ice to get the temperature, as we could not thrust the thermometer directly into the ice. The contents of the vessel must be stirred in order to keep the temperature everywhere the same. We can weigh the vessel before and after heating to see if there is any change in weight. The apparatus used is shown in Fig. 70. A tin pail supported over the burner *B.B.* is filled with ice and water. The temperature is measured by the thermometer *T*, and the contents of the pail stirred by the paddle *P*.

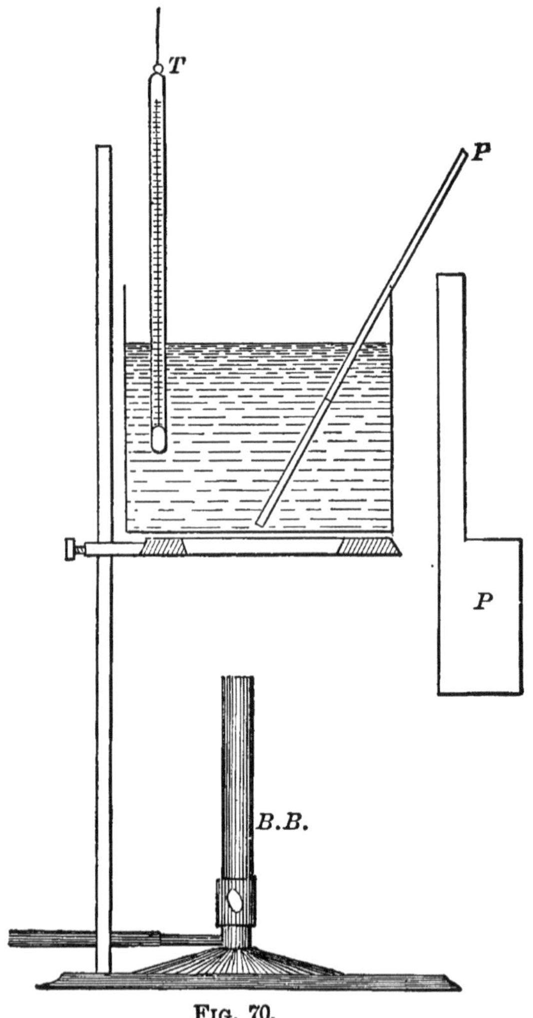

FIG. 70.

EXPERIMENT.

Apparatus.—Ring-stand and burner; tin pail; paddle; thermometer; ice or snow; water; watch or clock; spring-balance or rough scales; test-tube of cold water or plate of clean dry glass.

OBJECT.—To observe the effects of heating a body.

MANIPULATION.—Place about 100 cu. cm. of water in the pail; add enough ice or snow to fill the pail two-thirds full of the mixture; weigh the pail and contents on the spring-balance.* Record the weight. If ice is used, it should first be wrapped in a cloth and pounded until it is fine. By means of the paddle stir the contents of the pail vigorously until the thermometer reads zero. Adjust the thermometer so that the bulb is well covered, and does not touch the sides or bottom of the pail. Place under the pail a very low flame, turning the gas nearly off and almost closing the air holes in order to prevent backing-down. Stir vigorously, taking care that the solid and liquid are thoroughly mixed, and being careful not to break the bulb of the thermometer. Note the temperature at one-minute intervals until four minutes after the water has boiled. The readings must be continuous from the beginning to the end of the experiment. When the readings are completed, weigh again. From time to time during the work hold over the pail a piece of clean dry glass, or a test-tube filled with cold water, and observe the results. In addition, note *all that goes on,* and record all your observations.

Record results as follows:

Weight before =
Weight after =

TABLE I.

Time.	Temperature.	Remarks.

Under "Time," place the hour and minute of each reading. Under "Temperature," place the thermometer-reading to 0.1 degree. Under "Remarks," place any observations that you made at the time of the reading.

* A 64-oz. balance is the best.

Note carefully changes of form, appearance, formation of bubbles, moisture, etc.

QUESTIONS.—1. What effect has the addition of heat on the physical form of a solid? 2. What effect has the addition of heat on the physical form of a liquid? 3. Is it possible to add heat to a body and not raise its temperature? 4. Under what circumstances? 5. Under what circumstances does the addition of heat raise the temperature? 6. Is any change of weight produced? 7 Why is it necessary to stir the mixture? 8. When the temperature rises, is the rise regular?

GRAPHIC REPRESENTATION OF RESULTS. CURVE PLOTTING.

Where an experiment includes two sets of measurements, as the time and temperature measurements just made, the results are often expressed by means of a diagram. Suppose, for example, the following data have been obtained:

Time.	Temp.	Time.	Temp.
11 h. 3 m.	5°	11 h. 23 m.	4
5	7	25	3
7	4	27	2
9	3	29	4
11	3	31	5
13	5	33	6
15	8	35	5
17	6	37	4
19	7	39	3
21	5	41	0

To represent these results in a diagram, on a page of of your note-book, draw two lines at right angles, as *AB* and *BC* in Fig. 71. Let *BC* represent time and *AB* temperature. Divide the line *BC* into as many equal parts as there are observations recorded; in this case, 18.

Divide the line AB into as many equal parts as there are degrees between the highest and lowest temperature noted; in this case, 8. From each of the points a draw lines parallel to BC and equal to it in length, and from the points c on BC draw lines parallel to AB, thus dividing the paper up into a number of rectangles. Starting at the point where the two lines first drawn meet, mark on

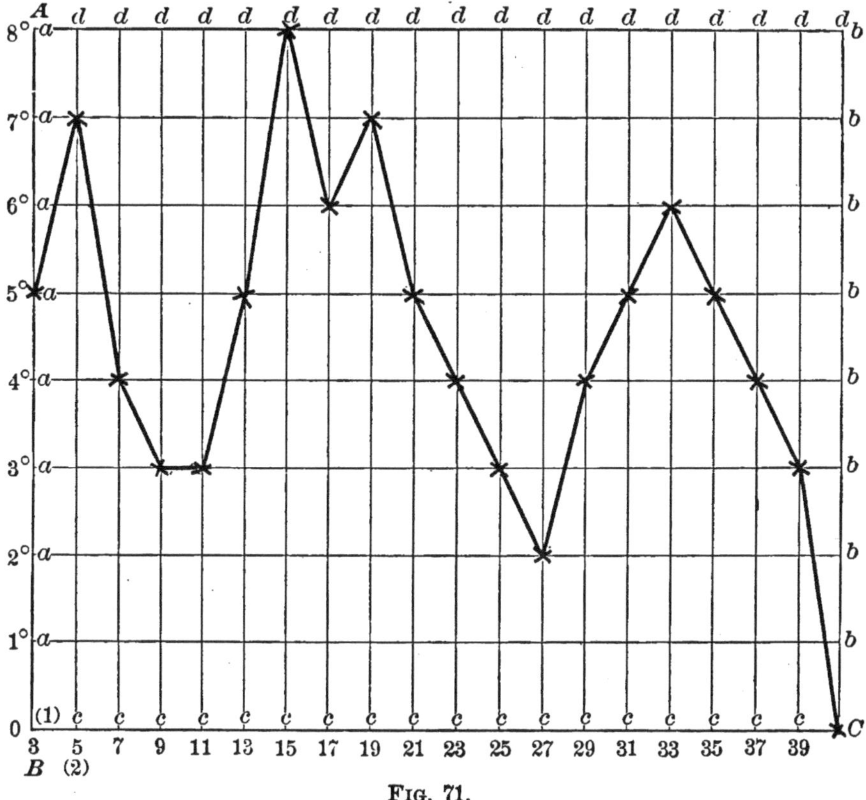

FIG. 71.

AB the temperature 5, the first observation, by making a cross five spaces up the line. Then on the next vertical line, which marks the time of the next observation, make a cross at the intersection opposite the next recorded temperature. Proceed in this way until all the temperatures have been entered. Connect the centres of the crosses by straight lines, or by a regular curve. The line so obtained will represent the temperature at various times. By inserting at the e laces on the curve anv observations that may be noted as regards changes, etc., a complete graphic account of the results of the experiment is obtained. This process is called *curve-plotting*, and is much used to represent to the eye the results of experiments.

Taking the results you obtained in the preceding experiment, plot a curve on one complete page of your note-book, representing the changes in temperature during the entire experiment. Enter at the proper points of the curve any phenomena that may have been observed. A curve like this is often called a *temperature-curve*. If BC be drawn the long way of the note-book page, the blue-ruled lines will be convenient for the vertical lines. The horizontal lines should be ruled very lightly with a pencil. Evidently any standard of length can be used in laying off the spaces. In Fig. 71, 1 cm. on the line BC represents two minutes of Time, and 2 cm. on the line AB represent one degree of Temperature if $AB = 16$ and $BC = 19$ cm.

EXERCISE 4.

LAWS OF COOLING.

EXPERIMENT.

Apparatus.—A small beaker-glass; thermometer, and clamp to support it; watch or clock; tin pail; water; ring-stand; and burner for heating the water.

OBJECT.—To observe the change in temperature in a cooling body.

MANIPULATION.—Place about 50 cu. cm. of hot water in a beaker glass, suspend the thermometer in the liquid, and observe the temperature every minute for twenty minutes. Every five minutes also observe the temperature of the room. Record results as follows:

Temp.	Time.	Temp. of Room.	Weight used.

QUESTIONS.—1. What have you observed regarding the changes in temperature when a body cools? 2. Does the quantity of the body make any difference? 3. Plot curves showing the changes in temperature in each case. 4. Does the difference in temperature between the body and the air seem to affect the rate at which the temperature falls?

EXERCISE 5.

MELTING AND BOILING POINTS.

Preliminary.—In the following exercise it is desired to observe the melting-points of some solids, and the boiling-points of some liquids. The melting-point of a solid is usually determined by immersing a small portion of it in some liquid having a sufficiently high boiling-point, heating the liquid, and noting its temperature when the solid melts. For bodies which it is supposed will melt below 100°, water is used; for higher melting-points, other liquids. The solid is usually placed in a small tube, called a *melting-tube*, which is attached alongside the thermometer. A melting-tube is made by drawing out a piece of glass tubing in the flame of a Bunsen burner, as shown in Fig. 72. The tube should be about 10–15 cm. from the narrow neck to the upper end, and the small part about 3 cm. long. Melt some of the solid (paraffine, for instance) in a small dish, draw a little up into the tube, and after wiping the outside allow it to cool. In determining the boiling-point of a liquid, a test-tube, containing about 1 cu. cm. of the liquid takes the place of the melting-tube.

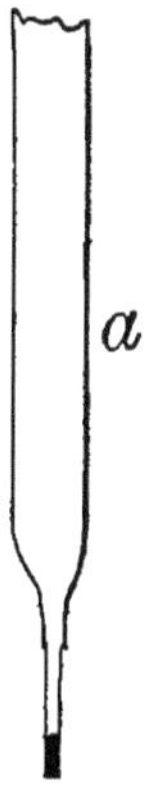

FIG. 72.

EXPERIMENT 1.

Apparatus.—Ring-stand; sand-bath and burner; thermometer, and some means of supporting it; "melting-tubes;" wax or paraffine in small dish (if tubes are unprepared); water; stirring-rod; rubber bands.

OBJECT.—To determine the melting-point of a solid.

MANIPULATION.—Attach the tube containing the solid

to the thermometer by elastic bands or strings, as indicated in Fig. 73, the solid being on a level with the thermometer bulb, and support the whole so that the bulb is about in the centre of the dish of water. Heat the water slowly, stirring gently, and note the temperature either when the solid becomes transparent, or when it slides up the tube. Either of these may be taken as indicating the melting-point. The second gives a little higher temperature than the first. Add some cold water, and repeat with a fresh tube. If possible, try more than one substance.

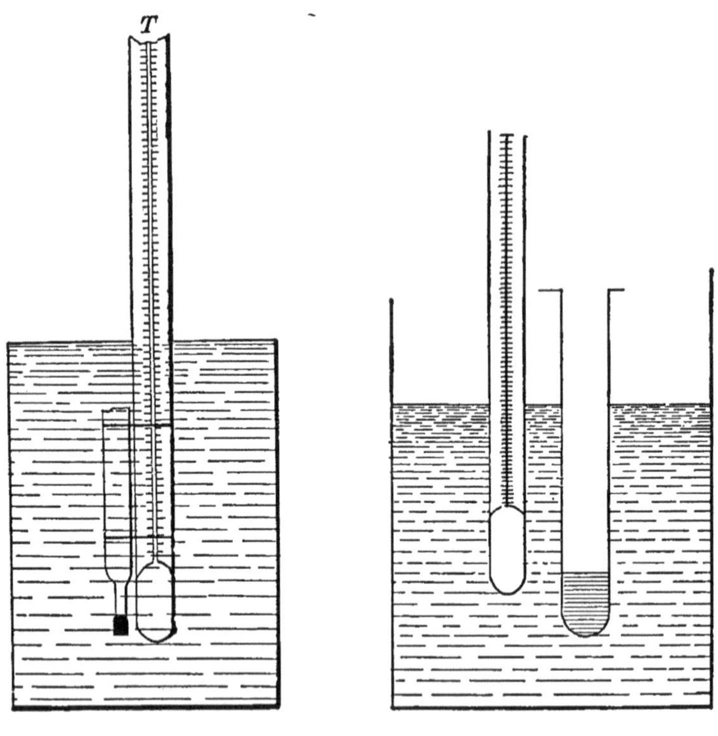

FIG. 73. FIG. 74.

EXPERIMENT 2.

OBJECT.—To determine the boiling-point of a liquid.

MANIPULATION.—Place about 1 cu. cm. of the liquid in a test-tube and support the tube in the beaker, which has been filled with cold water, so that the liquid is in the centre of the water, as in Fig. 74. Warm the water slowly, stirring gently with the thermometer, and note the temperature at which the liquid begins to boil. It is best to shake the tube gently while heating. Make two determinations with one liquid, and, if possible, repeat with another liquid. *Alcohol, ether, etc., are very inflammable, and should be kept away from the fire.*

EXERCISE 6.

HEAT CAPACITY.

Preliminary.—In the following exercise we wish to study the conditions affecting the rise in temperature of bodies when heated. We already know that the longer a body is heated, the more its temperature rises,* provided no change of form takes place, and that the greater the difference in temperature between the source of heat and the body, the more rapid will be the change.† In the following exercise we wish to see if the nature and quantity of the body heated have any effect.‡

EXPERIMENT.

Apparatus.—Part I: Test-tube, with perforated cork; thermometer; tin pail; ring-stand; burner; water; scales and weights; a burette or graduated cylinder.

Part II.: Three test-tubes with corks; mercury; alcohol; empty tumbler; also the apparatus for Part I.

OBJECT.—To observe the effect on the rise of temperature in a body produced by (*a*) quantity, (*b*) material.

MANIPULATION.—*Part I.* Effect of Quantity:—Place 4 grams of water in the test-tube, close the mouth of the tube by a cork through which the thermometer passes, the bulb being immersed in the liquid. Have ready the large vessel in which the water is boiling, place the tube in the water, stir it gently, and observe the time required for the thermom-

---

* Ex. 3. † Ex. 4.

‡ Let each pupil prepare a statement of the conditions under which such an exercise must be conducted.

eter to rise 10 degrees. Repeat the experiment with twice the weight of water.

Part II. Effect of Material:—Weigh out in three tubes 15 grams, respectively, of water, mercury, and alcohol. Cork the tubes, and stand them upright in an empty tumbler until ready for use. Take the tube containing the water and substitute for its cork the one bearing the thermometer. Having the thermometer bulb immersed in the water, read the temperature of the liquid and then plunge it into the boiling water. Shake it gently, and observe the time required for the thermometer to rise 2 degrees. Repeat the experiment with the other liquids. Tabulate the results as follows:

Wt. used.	Original Temp.	Final Temp.	Time of Immersion.	Body.

QUESTIONS.—1. For a given substance, can you make out any relation between the rise of temperature and the weight used? 2. Will exposure to the same temperature for the same time cause the same rise in temperature in all bodies?

EXERCISE 7.

DETERMINATION OF SPECIFIC HEAT.

Preliminary.—We observed in the preceding experiment that when equal weights of various bodies were exposed to the same temperature for the same time, the temperature of some rose more rapidly than that of others. This is indicated by saying that they have different *heat capacities.* The lower the heat capacity, the higher the temperature rose; and the less the temperature rose, the greater the heat capacity. Which has the greater heat capacity—water or mercury?

The unit of heat quantity is the heat required to raise a unit of weight of water from 0 to 1° on a temperature scale; hence the quantity of heat taken by any quantity of water is found by multiplying the weight of the water by its rise in temperature, or

$$H = W \times \text{Rise in temp.}$$

Evidently we can have several units of heat, according to the unit of weight taken and the thermometer scale used. The units of weight generally used are metric, the kilogram and the gram, and the scale used is the centigrade. The corresponding unit of heat is called the *calorie*. When kilograms are used the calorie is designated by a capital C; when grams are used, by a small c.

In order to be able to use these heat-units for all bodies, we must compare their heat capacities with that of water, which is taken as the standard. This comparison is expressed as the ratio of the heat capacity of the body to that of water, and this ratio is called the *Specific Heat* of the body. As the specific gravity of a body expresses how many times its density is that of water, so the specific heat of a body expresses how many times its heat capacity is that of water.

The simplest way to determine specific heat would be to take equal weights of water and the body to be tested, expose them for the same length of time to the same temperature, and note the rise of temperature. The higher the temperature of the body rose, the less would be its heat capacity. If, for example, the temperature of the body rose twice as much as that of the water, its capacity would be half as much, and its specific heat one half; if it rose half as much, its specific heat would be 2, etc. Calculate from the data of the preceding exercise the specific heat of mercury.

Specific heat is usually determined by what is called the "method of mixture." This consists in heating the body to a known temperature, and then bringing it in contact with water in a vessel, called a *calorimeter*, which will not lose heat by radiation. The body will lose heat and the water will gain it, until both are at the same temperature. Then the quantity of heat gained by the water will equal the quantity of heat lost by the body. If we know the weight of water in the calorimeter, and how many degrees it was raised, we can calculate how much heat was gained by the water, or, the equivalent, that lost by the body. If we also know the weight of the body, and the number of degrees it cooled, we can determine how many heat-units would be given out by one gram of the body cooled 1°. Dividing this by the amount of heat that would be given off by one gram of water cooled 1°, we can get the specific heat. In the following exercise we use an iron ball of known weight, heated in boiling water to 100°. Taking a known weight of water and noting the rise in temperature, we calculate the specific heat.

EXPERIMENT.

Apparatus.—Body whose specific heat is to be determined. Tin pail; ring-stand and burner; water; thermometer with white card; scales and weights; calorimeter; three corks; thread.

OBJECT.—To determine the specific heat of a given solid.

MANIPULATION.—Fill the pail about three-fourths full of water, and light the burner under it. Weigh the calorimeter, fill it about two-thirds full of water,* and weigh

---

* The point is not to get so much water in the calorimeter that it will overflow when the body is put in, but yet to have enough to completely cover it. A preliminary trial may be needed. Place the body in the calorimeter, and then add water enough to fill the calorimeter to about 5 cm. of the top. On withdrawing the body, the amount of water for good working conditions remains, and may be weighed.

again. Support the calorimeter on the corks upon the table at some distance from the heating apparatus, and put the thermometer into it. Weigh the solid under examination. When the water in the pail boils, suspend the solid in the centre by means of a thread, and allow it to remain there until it has assumed the temperature of the water (100° approximately). This will take about four minutes. Then read the thermometer and quickly transfer the solid to the calorimeter. Stir gently with the thermometer and read it to 0.2 at intervals of half a minute (record these readings) until the temperature of the water begins to fall. The highest reading of the thermometer is the temperature to which the solid heats the water. If time allows, repeat with different quantities of water in the calorimeter. Record results as follows:

Weight of body =
Calorimeter + water =
Calorimeter alone =

Water alone =

Temperature of body before immersion =
" " water in cal. after body is in =
" " " " " before body was in =

Increase in temp. of water in calorimeter =
Fall in temp. of body =

QUESTIONS.—Explain why the space between the walls of the calorimeter is filled with excelsior. Suggest some other substances that would do as well. Why should the solid be suspended by a thread rather than by a wire? What error is introduced by doing so? What error is introduced in calling the original temperature of the ball 100°?

CALCULATION.—Call the weight of the body W' and the weight of water in the calorimeter W, the original temperature of the water t and the final temperature t'. Then W' grammes of the body in cooling from 100° to t' degrees gave off sufficient heat to raise W grammes of water from t degrees to t' degrees, or raised it $t' - t$ degrees; the amount of heat given up was

$$W' \times t' - t.$$

As the heat was given up by W grammes of the body cooled from 100° to $t - t'$ degrees, the amount of heat given out by *one* gram cooled *one degree* was

$$\frac{W \times t' - t}{W \times 100 - t'}.$$

Since this also represents the amount of heat that would be required to raise one gram of the body 1°, we can get the *specific heat* of the body by dividing this value by the amount of heat *that would be required to raise* 1 *gram of water* 1° or 1c; so

$$\text{Specific heat} = \frac{\dfrac{W \times t' - t}{W' \times 100 - t'}}{1c}.$$

By substituting in this equation the values obtained in the experiment, the specific heat of the body may be calculated.

In this calculation we have neglected to take into account the heat that went into the calorimeter, which was itself heated. To make this correction, weigh the vessel that formed the inside of the calorimeter.* This gives the weight which was heated from the original temperature of the water to its final temperature, or $t' - t$; and so the heat that went into the calorimeter = Wt. cal. $\times\ t' - t \times$ the specific heat of the material of the calorimeter.†

* Or, if a metallic calorimeter be used, weigh the whole vessel.

† Glass, 0.198; brass, 0.858; iron, 0.1124.

The *total* heat given out by the body equals that which went into the water plus that which went into the calorimeter. The corrected formula, then, would be:

$$\text{SPECIFIC HEAT} = \frac{(W \times t' - t) + (\text{wt. cal.} \times t' - t \times \text{sp. heat of cal.})}{\dfrac{W' \times (100 - t')}{1c}}.$$

Re-calculate your value for specific heat with the correction for the calorimeter.

EXERCISE 8.

LATENT HEAT.

Preliminary.—The name *latent heat* is given to the heat which is required to change a solid to a liquid, or a liquid to a gas, without altering its temperature. Latent heat is usually determined (1) by measuring the heat given out by a known weight of vapor at the boiling-point of the liquid when condensed *at that point*, or (2) by determining the heat absorbed by a known weight of the solid at the melting-point when changed to a liquid at that temperature. In this exercise, the number of heat-units given off by one gram of steam at 100° when changed to water at that temperature is to be ascertained. The steam is condensed under such conditions that the heat given off will raise the temperature of a known weight of water. If we know the weight of steam condensed, the weight of water heated, and the number of degrees that it was raised by the condensed steam, we can calculate the latent heat.

The apparatus used in the first method consists of a glass vessel *D*, Fig. 75, holding the water, and containing a glass coil *E*, which terminates in a tube that projects from the bottom of the vessel. This coil is connected by the tube *C* with the flask *A*, which furnishes steam. *K* is an arrangement to prevent any condensed steam from passing into the coil. The tube *C* is thickly wound with cloth to stop any loss of heat and resulting condensation of steam before reaching *E*. When steam from *A* reaches *E* it condenses,

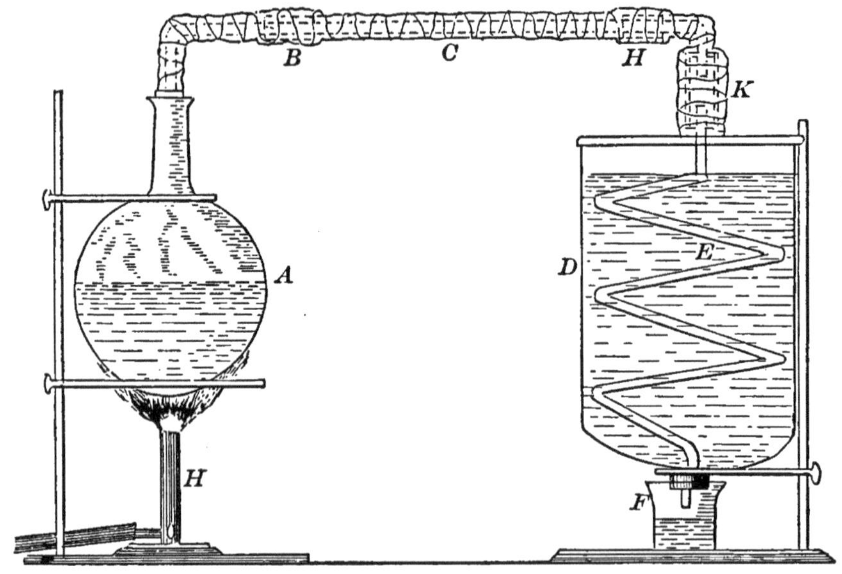

FIG. 75.

and so raises the temperature of the water in D. The condensed steam runs out into a vessel, and the weight of water so formed gives us the weight of steam condensed.

EXPERIMENT 1.

Apparatus.—Ring-stand; wire-gauze; burner; small beaker-glass; scales and weights; thermometer; liter and half-liter flasks; paddle; apparatus as in Fig. 75.

OBJECT.—To determine the number of heat-units required to change one gram of steam at 100° to water at 100°.

MANIPULATION.—Fill the flask *A* about two-thirds full of water and set it to boiling, the delivery-tube being disconnected at *H*. Place a measured quantity of water in D* (2 to 4 liters, according to size), and suspend the ther-

* The object, of course, is to get a known weight of water in *D*. This could be done by weighing *D*, filling it with water, and weighing it again. It is much more convenient, however, to measure the

mometer in the liquid. As 1 cu. cm. of water = 1 gram, you now know the weight of water in D. Weigh the small glass vessel *F*. When "live steam" comes out of the delivery-tube at *H*, connect it with the glass coil and allow it to run for three minutes, meantime stirring the water in D with the paddle.* Read the temperature of the water in D, and immediately place the small glass vessel under the end of the glass coil to catch the distilled water. Let the apparatus run from 15 to 20 minutes, stirring the water in D gently all the time; then remove the glass vessel containing the condensed steam, and immediately read the temperature of the water in D. Disconnect the steam-pipe and turn out the gas under *A*. Weigh the vessel containing the condensed steam. Arrange the results as follows:

Water in the condenser =
Original temperature =
Final temperature =

Rise in temperature =

Vessel + condensed steam =
Vessel alone =

Weight of steam condensed =

Calculation.—Call *W* the weight of water in the condenser, *W'* the weight of steam condensed, *t* the original temperature of the water in D, and *t'* the temperature to which it was raised. Then $W \times (t' - t)$ = the heat gained

water. It is best to fill *D* within about 5 cm. of the top, and it is convenient to take a volume which can be measured by liter or half-liter flasks.

* In stirring, care should be used not to break the glass coil, and to continually stir the water up from the bottom, at which point the colder water will always collect. The thorough stirring of the water is an important point all through the experiment.

by the water in D during the experiment. This amount of heat includes not only the heat given out by the steam at 100° when condensed to water at 100°, but also the heat given out by the water so formed in cooling from 100° to the temperature of the water in D, which varied from t degrees at the beginning to t' degrees at the end of the experiment. The *average* temperature to which this water was cooled was $\frac{t'-t}{2}$, and the amount of heat given off by the water formed from the steam in cooling was *

$$W' \times \left(100 - \frac{t'-t}{2}\right).$$

So that the heat gained due to the condensation of steam *alone* is

$$W \times (t' - t) - \left[W' \times \left(100 - \frac{t'-t}{2}\right)\right],$$

and the amount of heat per gram is

$$\frac{W \times (t' - t)\, W' \times 100 - \frac{t'-t}{2}}{W'}.$$

There are two errors that still need correcting for, if the results are to be at all accurate. (1) The glass vessel D and the glass coil E are raised from t degrees to t' degrees, as well as the water inside; hence the value taken for the heat given out is too small by the amount that went into the glass. To correct for this, weigh D and the coil together. Calling this weight G, the heat that went into the

---

* The original temperature being 100°, the average *fall* in temperature would be $100° - \frac{t'-t}{2}$.

glass is $G \times (t' - t) \times$ specific heat of glass.* If this quantity be added to the value taken before for the heat given up to **D**, it will give more accurate results. (2) During the experiment the water in **D** was losing heat to the air in the room.† This error can be avoided by filling **D** with water a number of degrees below the temperature of the room,‡ and stopping the experiment when it is heated as much above the temperature of the air as it started below. Then the water in **D** takes as much heat from the air while below its temperature as it gains while above, and so this error cancels out. The corrected calculation, then, is

LATENT HEAT =

$$\frac{[G \times (t'-t) \times \text{sp. heat steam}] + [W \times (t'-t)] - \left[W \times \left(100 - \frac{t'-t}{2}\right)\right]}{W'}.$$

SUBSTITUTE EXPERIMENT.

Preliminary.—The apparatus is shown in Fig. 76. Steam is generated in the flask, and passes through the covered tube into the calorimeter containing a known weight of water. The temperature of this water is observed before and after running in steam, and the increase in weight of the calorimeter gives the weight of steam condensed. The tube is covered with cloth to prevent condensation. The calorimeter is supported on a block or box, so that by removing the support, the calorimeter can be quickly

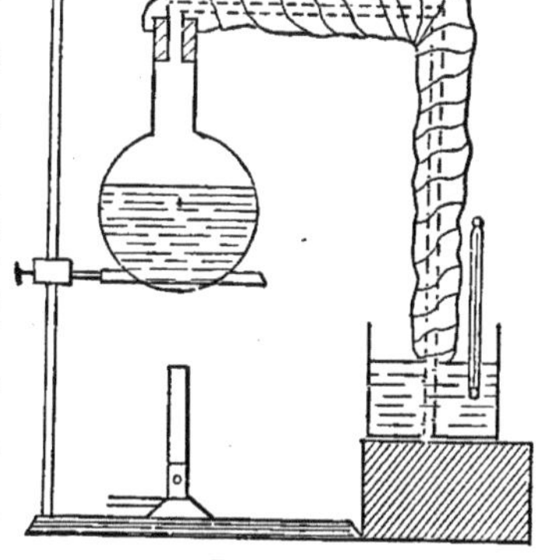

FIG. 76.

* This may be taken as 0.198.

† For another method of correcting for this error, see Worthington, p. 190.

‡ This can usually be done by taking water from the faucet after letting it run awhile, where there is a water service; otherwise, a little ice or snow may be added.

dropped down so as to clear the end of the delivery-tube without disturbing the flask.

Apparatus.—Ring-stand; wire-gauze; burner; flask with cork and delivery-tube; scales and weights; calorimeter and thermometer with white card; support for calorimeter (books, block of wood or small box).

OBJECT.—To determine the amount of heat given out by one gram of steam at 100° in condensing to water at 100°.

MANIPULATION.—Weigh the vessel to be used as a calorimeter, add about 300 cu. cm. of water, weigh again, and place the thermometer in the vessel. Fill the flask two-thirds full of water and heat. After the steam has escaped freely from the delivery-tube for three or four minutes, note the temperature of the water in the calorimeter, and as quickly as possible plunge the delivery-tube nearly to the bottom of it. While the steam is condensing, stir gently with the thermometer, noticing the temperature from time to time. When the temperature of the water in the calorimeter is 8° or 10° above that of the room, withdraw the delivery-tube as rapidly as possible, and remove the lamp from beneath the flask. Note the temperature of the water in the calorimeter, and weigh the latter. Record the results as follows:

Calorimeter + water	=	Steam condensed	=
"	=	Temp. before	=
Water	=	Temp. after	=
Calorimeter + water after	=	Gain in temp.	=
Calorimeter + " before	=		

EXERCISE 9.

COEFFICIENT OF LINEAR EXPANSION.

Preliminary.—We know that when a body is heated it expands. The fraction of its length at 0 that it expands when heated 1° is called its *coefficient of linear expansion,* or *the linear coefficient of expansion.* The fraction of its bulk at 0 that it expands when heated 1° is called its *coefficient of cubical expansion,* or *the cubical coefficient of expansion.* Evidently fluids would have no coefficient of linear expansion, but solids would have both.* In the following exercise we wish to determine the average linear coefficient of a metallic rod.

First Method. The apparatus is shown in Fig. 77. The rod *c* is surrounded by a large tube *ee,* which is first filled with ice-water and then with steam, thus heating the rod

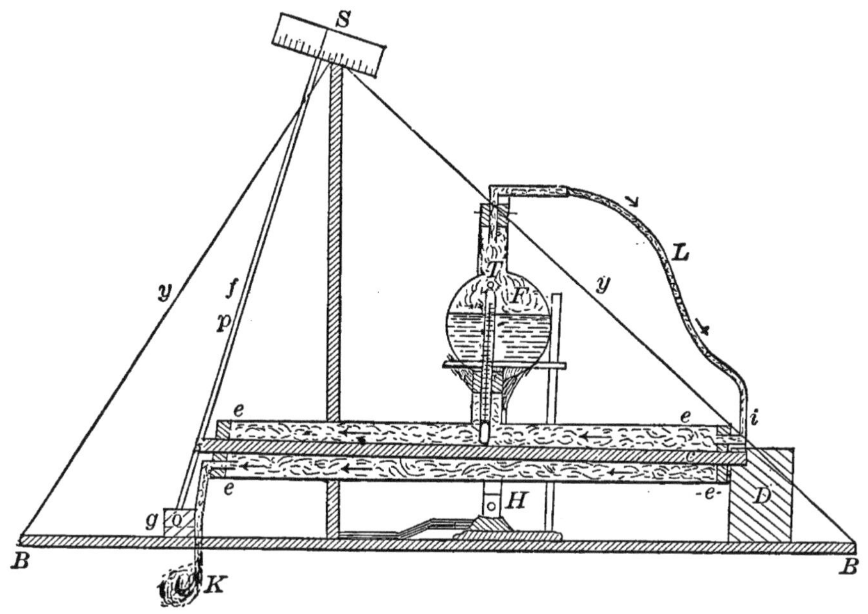

Fig. 77.

100°. The changes in length are magnified by the lever *p,* which reads on the scale *S.* In this way we can measure the change in length of the rod when heated from 0 to 100° and determine the required coefficient.

---

* Suggestion: Give the principles on which such measurements would be based.

EXPERIMENT 1.

Apparatus.—As shown in Fig. 77. Also, water; ice; funnel; tin pail; meter-stick; flask; ring-stand; burner; connecting tubes.

OBJECT.—To determine the linear coefficient of expansion of a solid.

MANIPULATION.—The chief error in this experiment is in the determination of length, since only the part of the rod inside the corks is at exactly the measured temperatures, though the rest of it becomes heated by conduction and expands to some extent. If we take the length between the outsides of the corks, we shall come very near the truth.

To get the length at zero. Attach the glass funnel to the rubber tube *L*, place a vessel under the exit-tube *K*, and pour ice-water through the apparatus until the thermometer *T* has read zero for several minutes. Read carefully the position of the pointer on the scale, and measure the length of the rod between the outsides of the corks. Allow the water to run out. Connect with the flask *F*, and run steam through the apparatus until the pointer again comes to rest; then note its position. If time allows, again pour ice-water through, and see if the pointer comes back to its first position. If so, repeat the experiment; if not, record the fact.

To find increase in length. By means of compasses, measure carefully the distance between the two pivots, repeating several times, and recording each measurement. Measure also the distance from the *lower* pivot to the point on the needle at which you took your reading. The average length of the "long arm" divided by the average length of the "short arm" gives the magnifying-power of the pointer. Record results as follows:

Length of rod =
Mag.-power of the pointer =

TABLE OF MEASUREMENTS.

	1st Trial.	2d.	3d.	Av.
Short arm,				
Long arm,				
Magnifying-power =				
Reading pointer at 0 =				
" " " 100 =				
Increase on pointer =				
True increase =				
Coefficient =				

CALCULATION.—Substract the smaller reading of the pointer from the larger; this gives the space traversed by the pointer. This, divided by the magnifying-power of the pointer, gives the true increase in the length of the rod. This increase, divided by the length of the rod at zero, gives the coefficient of linear expansion as a decimal which is to be carried out to the fourth place of significant figures. Or, calling L the length of rod at 0, M the magnifying-power of the pointer, S the reading of the pointer at 0, S' the reading of the pointer at 100,

$$\text{Increase, } 100° = \frac{S' - S}{M};$$

$$\text{Increase, } 1° = \frac{S' - S}{M \times 100};$$

$$\text{Linear coefficient of expansion} = \frac{\frac{S' - S}{M \times 100}}{L}.$$

Second Method. The apparatus used is shown in Fig. 78. The rod is inside the tin jacket R, which can be filled with either steam or ice-water, thus changing the temperature of the rod 100°. For measuring the increase in length, the pointer *P*, moving on the clock face, is attached to a screw, D, which advances a known distance for each turn. The end of this screw is in line with the end of the rod, and the screw is connected with one wire

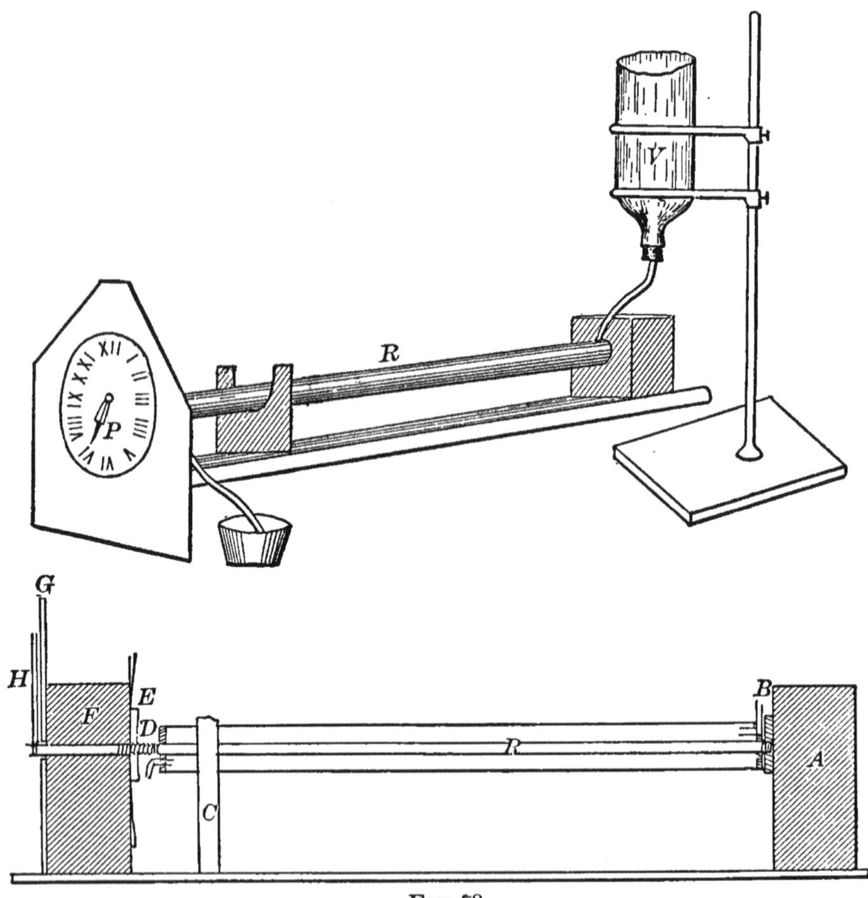

Fig. 78.

leading from a source of electricity, while the rod is connected with the other wire. When the end of the screw touches the end of the rod the circuit is completed, and this fact is indicated by some instrument placed in the circuit. If we know how far the screw advances at one turn, the change in length of the rod may be very accurately measured by finding how many turns of the screw move its end enough to make contact with the end of the rod first at 0 and then at 100°.

EXPERIMENT 2.

Apparatus.—As shown in Fig. 78. Also, ice-water; tumbler; meter-stick; boiler, burner, and ring stand; current of electricity and some instrument to indicate when the circuit is completed (galvanometer; lamp; sounder).

OBJECT.—To determine the linear coefficient of expansion of a metallic rod.

MANIPULATION.—Arrange the apparatus as in Fig. 78, connecting the glass vessel *V* with the jacket *R*, as shown. Turn the pointer, *P*, until the end of the screw, *D*, is just in contact with the end of the rod. Place a vessel* under the end of the exit-tube, fill *V* with ice-water, and allow it to run through the jacket, thus surrounding the rod with water at 0. As the rod contracts, turn the pointer so as to just keep contact between the end of the screw and the end of the rod, and continue until there is no further change in length. During this time keep ice-water constantly running through the jacket. When the rod has ceased to contract, read the position of the pointer at which the end of the screw just touches the end of the rod, and turn the pointer back half a revolution or so. Replace *V* by a flask two-thirds full of water, and connect it with the jacket *R*. With a Bunsen burner heat the flask and pass the steam through the jacket. During this operation turn the pointer back as fast as contact is made by the expansion of the rod. The expansion will be very rapid. When the rod has ceased to expand, note the position of the pointer at which contact is just made, and record the total number of minutes on the

* When this vessel is full, the contents are to be poured back into *V*. It is well to have a little ice in it to hold the temperature of the water at 0.

clock face which the pointer was turned back from its position when the rod was at 0. This number, multiplied by the decimal of a millimeter which the end of the screw moves for 1 minute on the scale, gives the increase in length of the rod when heated from 0 to 100°. Again attach the glass vessel *V*, run in ice-water, cool the rod to 0, and ascertain the number of minutes on the scale which the pointer must be turned to get contact again. These numbers should be very nearly the same. Average them, and calculate the coefficient of linear expansion of the rod for 10 cu. cm., taking for the length of the rod the distance between the outer ends of the jacket corks. Arrange results as follows:

Reading pointer at 0 =		min.
" " " 100 =	____	"
Pointer moved		"
Reading pointer at 100 =	____	"
" " " 0		"
Pointer moved in cooling		

Average =

CALCULATION.—Call the length of the rod *L*, the number of minutes that the pointer moved *m*, and the distance screw moved for 1 m., *a*.

$$\text{Then the increase for } 100^\circ = m \times a,$$

$$\text{and the coefficient for } 100^\circ = \frac{m \times a}{L},$$

$$\text{and the coefficient for } 1^\circ = \frac{\frac{m \times a}{L}}{100}.$$

EXERCISE 10.

CUBICAL COEFFICIENT OF A LIQUID.

EXPERIMENT.

Apparatus.—Alcohol of known specific gravity ; test-tube with perforated cork, and glass tube, with scale; tin pail; ice-water; ring-stand ; lamp ; thermometer.

OBJECT.—To determine the cubical coefficient of expansion of alcohol.

MANIPULATION.—Put some ice and water in the tin pail, and while the mixture is cooling weigh the test-tube with the cork and small tube. Fill the test-tube nearly full of alcohol and crowd the stopper in tight, thus forcing a column of alcohol up the small tube. This column should not be over 4 or 5 cm. high, and no air-bubbles should remain in the test-tube. Weigh again. The increase in weight represents the weight of alcohol in the apparatus. The volume is found by dividing this weight by the specific gravity of the alcohol (marked on the bottle from which it was taken). Immerse the test-tube in the ice-water, allow it to remain there until the alcohol column has come to rest, and note the position of the top of the column on the scale. Remove the ice and slowly heat the contents of the pail, stirring gently with the thermometer until the alcohol column has risen four or five cm. Stop heating, and note when the alcohol column ceases to rise. Read its position on the scale, and at the same time note the temperature of the water. You have now the distance which the alcohol rose in the tube when heated from 0 to the final temperature of the water in the pail.

CALCULATION.—To determine the increase in volume, multiply the rise on the scale by the volume corresponding to a rise of 1 cm. as given on the card attached to the instrument. This gives the increase in volume. Record results as follows :

Weight of apparatus + alcohol	=
" " " empty	=
" " alcohol	=
Volume of alcohol	=
Alcohol column read at 0	=
" " " " t	=
No. of degrees alcohol was heated	=

From the data, calculate the cubical coefficient of alcohol for 1°.

EXERCISE 11.

COEFFICIENT OF EXPANSION OF A GAS AT CONSTANT PRESSURE.

EXPERIMENT 1.

Apparatus.—The special form in Fig. 79 ; ice or snow-water ; tin pail ; ring-stand ; thermometer ; meter-stick. If apparatus is not calibrated there will be required in addition : scales and weights ; mercury ; small vessel for weighing mercury ; burette or balances that can weigh 300 to 400 grams.

OBJECT.—To determine the coefficient of a gas under constant pressure.

MANIPULATION.—To find the volume of the gas used. Having no water in *B*, Fig. 79, remove the gas-holder *A*, and by means of a burette determine its volume in cu. cm.* Dry, and place inside *B*, as shown, and connect with the tube *DC* when full. The volume of gas under test is really that in *A* plus a little in the tube ; but since the cork occupies some room in *A*, it will be near enough to the truth to call the volume that of *A* alone. Fill *B* with water to about 3 in. above the level of *A* ; add some ice or snow. In a short time the temperature of the liquid should be 0°. While it is cooling, find the volume represented by 1 cm. on the tube *DC*. Detach the tube at *E*, and draw a column of mercury into it. Lay the tube on a

* Or, weigh *A* empty and then full of water. From the weight of water contained in *A*, calculate the volume.

scale and measure the length of the column. Call this length L. Weigh a small dish; pour into it the mercury from the tube; weigh the whole, and compute the weight of the mercury alone. Call this weight W. Then if Δ* = specific gravity of mercury,† the number of cu. cm. contained in length L is $\frac{W}{\Delta}$, and the volume in the tube, per cm. of length, $= \frac{\frac{W}{L}}{\Delta}$. Put this down, labelled "Volume per cm. in tube." It is best to repeat this several times and use the average values obtained.

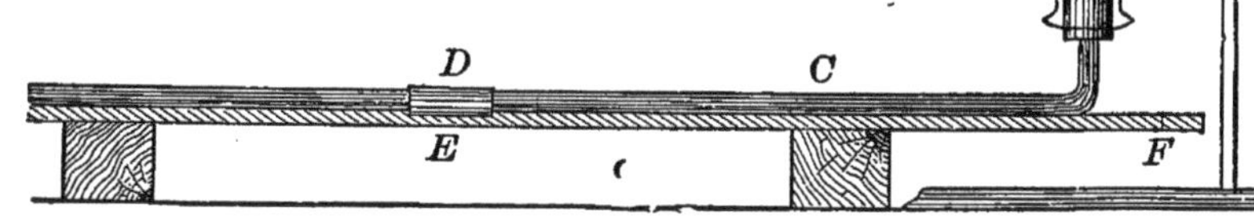

FIG. 79.

Next, draw a small globule of mercury about 1 cm. long‡ into the tube, and get it near E by gently inclining the tube and keeping the finger over one end. This is used as an index. Attach the tube at E, and read the position of the inner end of the index on the scale F. Before reading it is advisable to tap gently with a pencil on the tube over the index, as the index is liable to catch a little. If now the water in B is at 0°, add a little warm water to it, and stir thoroughly with the thermometer until a rise of one or two degrees in temperature takes place. As the gas in A increases in bulk the index will move out on the scale. When the index has assumed a constant position after tapping, and the thermometer is also constant (at t°), the dis-

* This sign is called *delta*.

† Say 13.6.

‡ The tube must be dry, and also the mercury. Read from the end of the meniscus.

tance moved by the index represents the increase in volume A of a cu. cm. of gas heated from 0° to $t°$. From these data calculate what decimal of its bulk at 0 a gas increases per degree centigrade. Put this down, carried out to the fourth decimal place, and label it "Coefficient of Expansion." Add ice to *B*, and repeat experiment several times with different temperatures ($t°$). Record results as follows:

Volume of gas taken =
Vessel weighed =

TABLE I.

Trial.	Length Mercury Col.	Wt. Mercury Col.	Vol.	Vol. per Cm.

Average volume per cm. =

TABLE II.

Index read at 0.	Index read at $t°$	Distance moved by Index.	Volume corresponding	Coefficient of Exp. for 1°.

Average value of coefficient =

A Second Method. The apparatus is shown in Fig. 80. The tube *k* is closed at one end, and contains a drop of mercury, *g*, to serve as an index. The air whose expansion is to be measured is contained between the closed end of the tube and the index. This tube passes through a cork in one end of a larger tube, *aa*, which is provided with an inlet tube *T′*, and an outlet tube *T*. The gas can be cooled to 0° by filling the large tube with ice-water, and heated to 100° by running in steam. By measuring the movements of the index, the coefficient may be calculated, as in the preceding experiment.

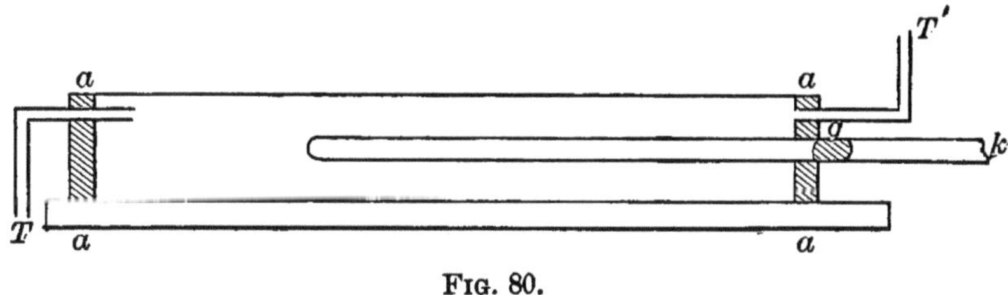

FIG. 80.

EXPERIMENT 2.

Apparatus.—Special form as shown in Fig. 80. Flask and connections; ring-stand; gauze and burner; ice- or snow-water; funnel, and vessel to hold ice-water; meter-stick.

OBJECT.—To determine the cubical coefficient of expansion of a gas at a constant pressure.

MANIPULATION.—Run steam into the large tube *aa*. As the index is forced out, push the tube in, keeping the inner end of the index just inside the outer edge of the cork. When the index remains stationary after tapping with a pencil, measure the distance from the inner end of the index to the outer end of the tube. Disconnect the steam, allow the apparatus to cool for a moment, and then connect the large tube with a funnel, and run ice-water through until the index again comes to rest, when kept just outside of the cork as above. After tapping again, measure the distance from inner end of the index to the outer end of the tube. Remove the tube and measure the distance from its open end to the inner side of the closed end. Record results as follows:

Dist. index from end of tube at 100°	=
" " " " " " " 0°	=
Index moved	=
Length of tube	=

The length of the tube minus the distance of the index at 0 from the open end gives the length of air-column used.

CALCULATION.—The length of the air-column at 0° represents the volume used, and the distance the index moved represents the increase for 100°; so

$$\text{Coefficient} = \frac{\text{Increase}}{\text{Vol. at } 0^\circ \times 100}.$$

EXERCISE 12.

ABSORPTION AND RADIATION.

Preliminary.—We know that when a heated body loses heat by radiation, bodies near it are warmed. In the following exercise we wish to study some of the conditions affecting the rate at which heat is absorbed by bodies when exposed to radiation. Under what conditions must tests be conducted? What conditions might affect the amount of heat absorbed by a body?

In the exercise we will use an iron ball to radiate heat, and keep it hot with a flame. To absorb the heat, we will use tin cans containing water. The cans are of the same size, but differ in character of surface—as rough or bright, in color, etc. The ball is suspended over the flame and the cans supported on blocks, as shown in Fig. 81.

EXPERIMENT.

Apparatus.—Part I: Ring-stand; iron ball and wire; two flat tin cans of the same size, one covered with lamp-black, each with a hole in the cover; blocks of wood to support the cans; two thermometers; watch or clock; water.

Part II: In addition to the above, tin pail for heating some water.

OBJECT.—To investigate the effect of color, character of surface, etc., on the amount of heat absorbed or radiated by a body.

Manipulation.—*Part I.* Suspend the ball by means of a wire about four inches above the Bunsen burner and light the burner. Fill each can two-thirds full of water, put on the covers, insert the thermometers through the holes, and support the cans at equal distances from the ball and on opposite sides, as in Fig. 81. Read the thermometers at intervals of a minute for four or five minutes. The cans are best placed with their largest flat sides towards the ball, and great care must be taken that the flame is equidistant between the two. Avoid any draught.

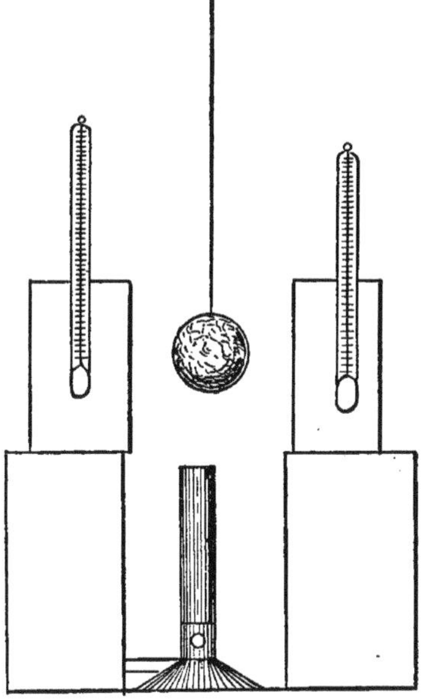

Fig. 81.

Part II. Fill both cans with equal amounts of water at about 10° above the temperature of the room. Read the thermometers again at intervals of a minute. Tabulate the results, and state what you have learned, as regards the effect of surface, color, etc., on absorption of heat when

(*a*) radiant heat strikes the surface;

(*b*) the body radiates heat.

State in your note-book three common examples of the application of these facts.

EXERCISE 13.

SOLUTION.

EXPERIMENT.

Apparatus.—Five test-tubes; water, and means of warming it; measuring-cylinder; scales and weights (if solids are not ready weighed); powdered and lump sugar; burner; sand; iodine; copper sulphate; alcohol.

Object.—To observe the conditions affecting solution.

MANIPULATION.—*Part I.* Take two test-tubes, place in each 0.5 gram of powdered sugar; to one add 5 and to the other 10 cu. cm. of warm water. Cork the tubes, and shake gently until all the sugar is dissolved; then add a second 0.5 gram to each, and so proceed as long as the solution goes on. Set the tubes aside. Can any amount of sugar be dissolved in a given amount of water? Does the volume of water used have any effect?

Part II. Place a piece of lump-sugar weighing about 0.5 gram in one tube, and an equal weight of powdered sugar in another. Put about 5 cu. cm. of warm water in each, shake gently, and observe the time required to dissolve. Does the condition of the body make any difference?

Part III. Warm one of the tubes containing the solutions formed in Part I. Then add another 0.5 gram sugar, and heat. Does the temperature of the water have any effect on its power to dissolve?

Part IV. Take five test-tubes, place in them about equal amounts (0.25 gram) of sugar, sand, iodine, copper sulphate, and alcohol. Add to each 5 cu. cm. of water. Are all bodies equally soluble? Can one liquid dissolve in another?

Part V. Repeat the experiment with alcohol. Does the nature of the liquid make any difference?

Part VI. Place about 40 cu. cm. warm water in a measuring-cylinder; read the volume; now put into the water 0.5 gram of powdered sugar, and again read the volume. When the sugar has dissolved, read the volume again. What effect on the volume of the liquid is produced by dissolving a liquid in it?

Tabulate all the conditions that you have found affected the solution.

DYNAMICS.

EXERCISE 1.

ACTION OF A FORCE UPON A BODY.

Preliminary.—In the following exercise we wish to study the behavior of a body when a force acts upon it. The force must be applied to a body that is free to move, or we cannot be sure that anything observed is due to the action of the force alone. If the body is suspended, the suspending wire takes its weight, and there is nothing to prevent its responding freely to forces applied horizontally. A string may be attached to the body, and pulled in various directions in a horizontal plane and with different degrees of force. The degree of force may be approximately, but not accurately, measured by attaching a spring-balance.

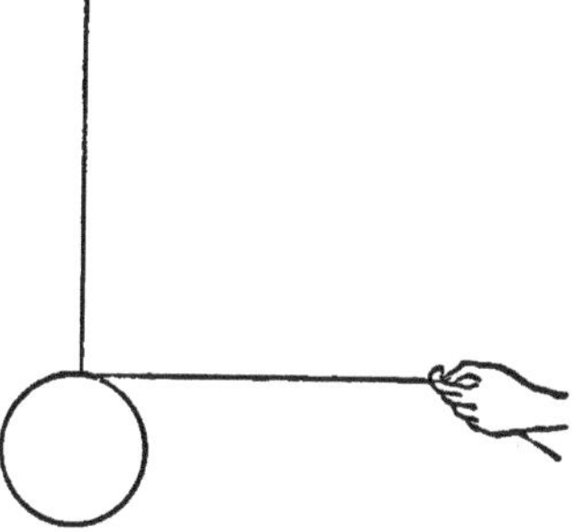

FIG. 82.

EXPERIMENT 1.

Apparatus.—Heavy body suspended by a wire; piece of cotton string; two spring-balances for Experiments 2 and 3.

OBJECT.—To see (1) what happens when a force acts on a body, and (2) the effect of the direction of the action.

MANIPULATION.—Arranging the apparatus as shown in Fig. 82, pull the string from various points of compass, always at the height of the ball and parallel with the table. Make five trials, and record the results in a table arranged as follows:

TABLE I.

Direction of Pull.	Action of Body.

Indicate the direction of the pull by inserting N. for north; W. for west; N.E. for north-east; etc.

EXPERIMENT 2.

OBJECT.—To see the effect produced by the magnitude of the force.

MANIPULATION.—Attach a spring-balance to the string. Pull suddenly on the balance, trying to vary the *amount* of the pull without altering the time during which it is applied. A sudden "yank" is best, only not hard enough to break the string. Observe approximately the amount of the body's motion and the strength of the pull as shown by the reading of the balance-index. Only general results are expected. Make four trials, and record the results as follows:

TABLE II.

Force.	Motion.	Direction.

In columns 1 and 2 insert the words, "more," "less," or "same," as the case may be. In the third column place the initials of the points of compass, as before.

QUESTIONS.—What is the inference in regard to the magnitude of the force? Does the direction in which the force acts make any difference in this respect? Make several trials in each direction, say three, making twelve trials in all. By studying the data obtained in both these experiments make a summary of what is indicated, as regards the action of a force on a body, in relation to (1) motion of the body, (2) direction of the motion as compared to the direction of the force, (3) amount of the motion as compared to the amount of the force. [This last in general terms only.]

EXPERIMENT 3.

OBJECT.—To observe the effects of equal and unequal forces acting in opposite directions.

MANIPULATION.—Attach two spring-balances by springs to the weight and pull in opposite directions. Observe the motion of the body when the forces are equal. By suddenly pulling stronger on one balance, render them unequal, and observe the results. Try several times, applying the two forces in various directions, but always opposite to each other and parallel to the table. Mark one force + and one − and record, in a table, as follows:

TABLE III.

+ Force.	− Force.	Result on Body.

Under each force place its value, obtained by reading the balance, and in the third column insert the words, "moved" or "no motion," as the case may be. In case the body moved in the direction of the + force, place the

+ sign before the word "moved"; if in the opposite direction, the − sign.

QUESTIONS.—1. Under what conditions does the body move? Under what conditions does it remain at rest? 2. Does the fact that a body is acted on by forces necessarily mean that the body will move? When a body is acted on by forces and the opposite forces are equal, it is said to be in *equilibrium*. Place in the note-books two cases of equilibrium that you know of, and explain in each case how the equilibrium is obtained.

EXERCISE 2.

THE FORCE OF FRICTION.

Preliminary.—When two surfaces are rubbed together some force is exerted. This fact is said to be due to *friction* of the surfaces, and the force with which the surfaces resist being rubbed is said to be due to the *force of friction*. In the following exercise we wish to study the conditions which affect the magnitude of this force. We must measure the force under various conditions. Since the force required to keep a body moving over a level surface is equal to the force of friction acting upon it, if we ascertain the magnitude of one force, we know that of the other. By turning the block used in the experiments edgeways or flat, we can vary the extent of surface rubbed; and by laying one or two blocks on the first, we can double or treble the weight without altering the extent of surface.

EXPERIMENT.

Apparatus.—Board; blocks of wood; 8-oz. balance or rubber-strip; string.

OBJECT.—To study the conditions affecting the magnitude of the force of friction.

MANIPULATION.—Lay the block of wood on its smooth,

flat side, and draw it along the board at a uniform rate of speed. Two students should work together on this experiment. One student, holding the balance horizontally on the palm of one hand, and grasping its ring with the other hand, should devote his whole attention to reading the balance as his hands slide along the board, with a view to determining the *average* position of the pointer. The other student should see that the motion is uniform and the pull parallel. Repeat several times, recording the force observed each time. Repeat with the block turned on edge. Lay the block flat and place a second one on it. Measure the force again. Try again with two blocks laid on the first. Repeat the first part with the rough side of the block down. Tabulate results as follows:

Position of Block.	Force.	Av. Force.	No. of Blocks.	Trial.

QUESTIONS.—1. What effect has the extent of surface on the force of friction? 2. What effect has the weight? 3. What effect has the character of the surface? 4. Weigh one block and calculate the coefficient of friction for all the weights, taking two blocks as twice the weight of one block, etc.

Measurement of Forces.—If forces are to be compared as to strength, we must have a unit of force, just as we had units of length, volume, etc. We need for this unit a force that can be readily obtained, and easily used for purposes of comparison. The force selected is gravitation, as shown in the pull of the earth on bodies upon its surface. This pull has been found to be always the same for the same body at the same place, but in order to get a definite pull we must also specify the quantity of matter to be pulled. The *unit of force* is taken as the pull of the earth on a unit weight—say a pound weight or a gram weight; thus a force of one pound would be a push or pull equal to the pull of the earth on a pound weight. In order to pull with a force of one pound, you would have to exert your muscles as much as in holding a pound weight. In the same way, a force of one gram would be a force equal to the pull of the earth on a gram weight.

PROBLEMS.—Explain what is meant by: 1. A force of 6 lbs.? 2. A force of 1 ounce? 3. A force of 10 grams? 4. A force of 1 ton? 5. A force of 6 kilograms? A pound weight weighs 453 grams. A force of 2 lbs. would be the force of how many grams?

There are two ways of measuring forces—by weights and by a spring-balance. In the first way, the weights are made to pull against the force to be measured, and a sufficient number of weights are used to just balance the force. The sum of the weights shows the value of the force. In the second way, the force to be measured is made to stretch the spring of the balance, and the value of the force is given by the index. A spring-balance when used for measuring force is sometimes called a *Dynamometer*.

Graphical Representation of Forces.—To represent a force whose magnitude, direction, and point of application are known, we proceed as follows:

1. Make a point for the point of application.

2. To that point rule a straight line whose *direction* is the direction of the force.

3. Assuming some particular length to represent a unit of force, with compasses or scale lay off this length along the line of direction as many times as there are units of force.

For example, suppose we have to represent a force of 6

lbs. magnitude acting in an upward direction on the line *ab*, Fig. 83. Mark the point of application anywhere on the line, say at *c*. To *c* draw a straight line *cd*. Starting at *c*, with a scale of 1 cm. to 1 lb., lay off along *cd* a distance of 6 cm. Suppose, again, we wish to represent another force of 3 lbs. acting at *e*. Draw to *e* a line *eg*, and from *e* lay off on it 3 cm. The direction of a force is sometimes represented by an arrow. The fact that forces are acting in opposite directions is also indicated by giving one the + sign and the other the — sign. Commonly, + is used for the upward direction and — for the downward, + toward the right and — toward the left, but since the general rule is to mark one force + and all forces whose general direction is opposite —, the direction + represents should be stated in each diagram. Forces are added by laying off on the line representing the forces of one sign a line representing the forces of opposite sign. Thus, to find by construction the sum of + 6 and — 4, lay off on the line + 6 a distance equal to — 4. The part of the length not used up by the — length is the required sum, + 2 in this case.

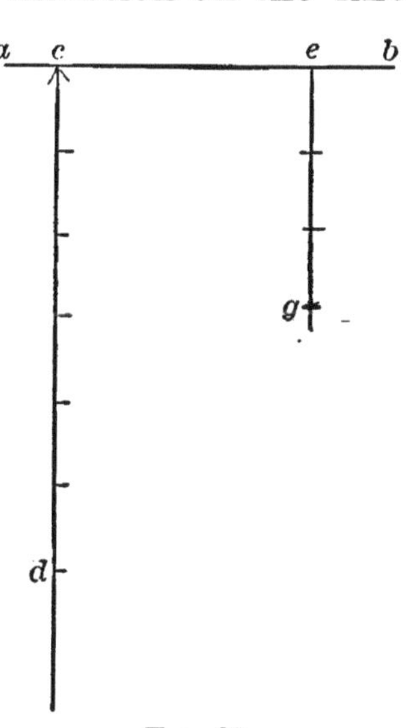

Fig. 83.

A problem can be worked out, as most convenient, by geometry, using lines as above, or by algebra, using numbers and signs. In the latter method the sign prefixed to the magnitude of a force indicates its direction relative to the other forces. These magnitudes are added and subtracted as other algebraic quantities. Thus the sum of two faces, one + 6 and — 4 is a force of + 2.

EXERCISE 3.

COMPOSITION OF FORCES.

Preliminary.—A force that in its action on a body is equal to the combined effects of several forces is called the *resultant* of those forces. For example, a single force which will produce the same effect on a horse-car as the force exerted by the two horses would be the resultant of the forces exerted by the horses. The process of finding the resultant of several forces is called the *composition* of forces. The reverse process, finding several forces whose combined effect on a body is equal to that of a given force, is called *resolving* that force, and the equivalent forces found are called its *components*.

We can apply more than one force to a body at one point in two ways.

(*a*) We can apply the forces in one straight line, in the same direction or in opposite directions, as a number of engines drawing a train, or two men pulling against each other on a rope. In the first case each force adds its effect to that of the others; in the second case, the lesser force diminishes the effect of the greater. The resultant of the forces in each case is their algebraic sum. For example, a wagon is pulled by three horses tandem, each exerting a force of 1000 lbs.; the same effect on the wagon would be caused by one horse exerting a force of 3000 lbs. Or, again, a body is acted upon by two opposite forces, $+ 8 - 3$, in strength; the same effect would be produced on the body by a single force $+ 5$ in strength.

(*b*) We can apply the forces at an angle to each other. Suppose two forces act on the same point, one pulling it east and one pulling it north. Can a single force be found that, when applied alone to the point, will produce the same effect as the two forces? To answer this question we must let two forces act on a body at an angle, and see if we can substitute for them one force that will produce the same effect. If we can do so, the substituted force will be the resultant.

Two balances, *A* and *B*, Fig. 84, are joined by a cord. On this cord slides another whose ends are attached to two

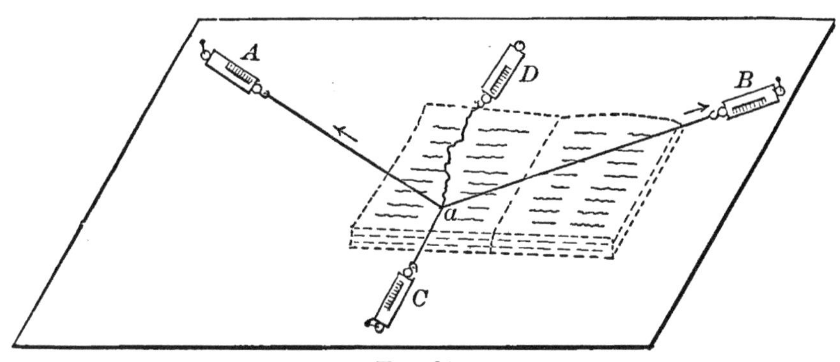

Fig. 84.

other balances, *C* and *D*. *A* and *B* are fastened to the table, and by pulling on *C* two forces will act on the point *a* at an angle *AaB*. The balance *C* will hold this point against the combined action of *A* and *B*, and this combined action will pull the index of *C* down to a certain point, and *a* will take a certain position which can be marked. Now if we release *A* and *B*, and by pulling on *D*, as in Fig. 85, can bring *a* to the same point, we shall have substituted the pull of a single balance for the combined pulls of the two balances, *A* and *B*. It will be also desirable to find out whether a single force acting on a point can be resolved into two components acting at an angle on the same point. We can do this by pulling on *a* with a certain force registered on *C*, and then trying to substitute angular pulls by *A* and *B* that will bring *a* to the same point.

EXPERIMENT 1.

Apparatus.—Four 24-lb. balances; fish-line; scale; dividers; nails; wooden block.

OBJECT.—When two forces act at an angle on the same point, to find one force whose effect on the point is the same as the combined effects of the other two.

MANIPULATION.—Arrange apparatus as in Fig. 84, *A* and *B* being fastened to the table. Pull on *C* in any direction until it and *A* and *B* read over five pounds. In a general way, the stronger the pulls the better, only care must be used not to break the strings. Either hold the balance *C* steady; or, better, secure it by a nail passing through the ring and driven into the table. Read the balances, and mark the position of the point *a* on the table. Pull on *D* with a force of a few pounds, and, still keeping a strain on D, *gently* release the balances *A* and *B*. Then by pulling on D, find the direction and magnitude of a force that will bring *a* to exactly the same point as when acted on by *A* and *B*. If you succeed in doing this, you will have one force producing the same effect on *a* as the combined forces *A* and *B*. Make several trials, and record the results as follows:

Force A.	Force B.	Force C.	Force D.

QUESTIONS.—1. How do *C* and D compare in magnitude? 2. How do they compare in direction? 3. How does the general direction of D compare with that of *A* and *B*? 4. How does the magnitude of D compare with those of *A* and of *B*? 5. How does the magnitude of D compare with the sum of the magnitudes of *A* and *C*?

EXPERIMENT 2.

Apparatus.—Four spring-balances; stout string; nails; etc.

OBJECT.—To find two forces that, if applied to one point at an angle to each other, will produce the same effect as a single given force applied to the same point.

MANIPULATION.—Arrange apparatus as in Fig. 85, *A* and *B* being loose and *C* being fastened to the table. Pull on **D** with any desired force in any direction, and note the readings of *C* and **D** and the position of *a*. Substitute for

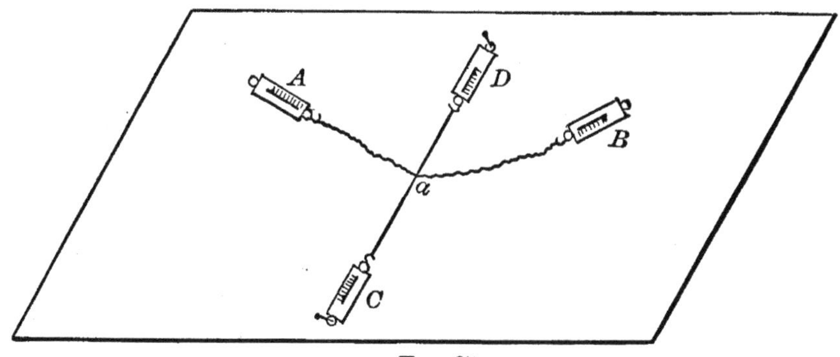

FIG. 85.

D, *A* and *B* pulled at an angle. See if their combined effect can be made to bring *a* to the point where it was when *C* and **D** alone acted on it. Make several trials with a view to finding out whether more than one combination of *A* and *B* can take the place of **D**. Record in each case the balance readings, and the angles made by *A* and *B* as estimated by the eye.

QUESTIONS.—1. Can two forces produce the same effect as one? 2. Can they be substituted for one? 3. How do their magnitudes compare with that of the one for which they are substituted? 4. Can more than one set of forces be substituted for the given force?

Parallelogram and Forces.—The next thing is to find a way to express graphically the relation of the resultant and its components. The readings of *A* and *B* in the apparatus represented in Fig. 84 will give the magnitudes, and the strings the directions of the components. Taking these directly from the apparatus, we can construct a parallelogram whose two adjacent sides represent the two components. Since, as we have just seen, the third force, *C*, is equal and opposite to the resultant, we can include the resultant in the parallelogram by taking the reading of *C* for its magnitude and making its direction exactly opposite to that of the string *Ca*. In the diagram we can look for any geometrical relation that may exist between component and resultant forces.

EXPERIMENT 3.

Apparatus.—Same as in Exp. 2, and a thick block of wood to be used as a ruler.

OBJECT.—When two forces act at an angle on the same point, to express graphically the relation between them and their resultant.

MANIPULATION.—Proceed as in Exp. 2. When all three balances are pulling on the point *a*, place the note-book under the strings so that the knot comes about in the centre of the page, Fig. 84. Bring the block of wood gently up to one of the strings so as to just touch it at all points, and, using it as a ruler, draw a line a few inches long. This line should represent exactly the direction of the string. Great care must be used not to move the string and not to let the block of wood touch the knots at either end. The harder the balances are pulling, the less the string is likely to be deflected. Do the same with the other two strings, using great care that the note-book does not change its position during the operation. Read each balance, and put its reading at the outer end of the corresponding line. Remove the note-book, and by means of a ruler prolong the lines until they meet at a point as indicated by the dotted lines in Fig. 86. This is the point, *a*, which was directly under the knot. You have now the magnitude, direction, and point of application of the three forces.* Starting from the point a, lay off on the lines

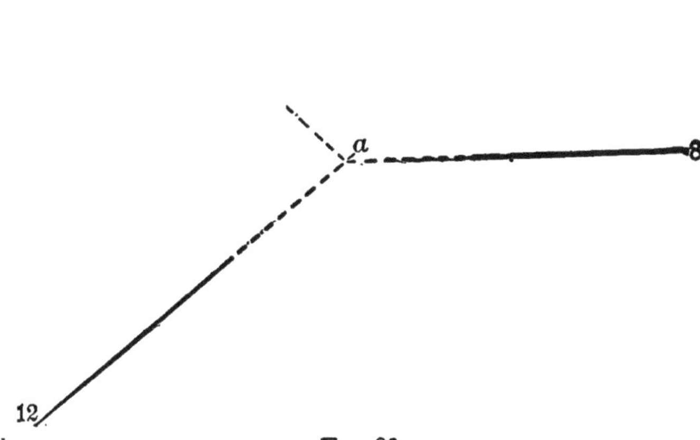

Fig. 86.

9 and 8 the magnitudes corresponding, representing 1 lb. by the largest unit of length the note-book page permits. Construct a parallelogram, of which the point a forms one angle and the lines representing the two forces 9 and 8 form the sides adjacent. The junction of the opposite sides is found by prolonging the line representing the third force, 12, as a diagonal through this parallelogram, and from a laying off on it a distance equal to that force. Make as many trials as time allows, using various forces. From the study of the diagrams so obtained make a rule for finding, by means of a diagram.

1. The resultant of two forces whose magnitudes and directions are known.
2. The components of a given force.

---

* Before laying off the forces, the balance-readings should be corrected for position error and zero error, if these errors are of magnitude. Usually, however, as they tend to counteract each other, the total error is so slight that it may be neglected. The record in the note-books should state whether or not corrected readings are used.

EXERCISE 4.

PARALLEL FORCES.

Preliminary.—So far as we have considered cases where the forces had the same point of application; but there is another case. Suppose the body *ab*, Fig. 87, to be acted on by two parallel forces, *x* and *y*. A good illustration of this is where a carriage is pulled by a span of horses. In this case it is evident that the two forces have different points of application, and that the resultant has a still different point. Forces applied in this way are called parallel forces. To investigate the case of parallel forces the apparatus shown in Fig. 88 is used. By pulling down on the balances *B* and *C*, parallel forces act on the meter-stick, and the combined pull of the two is taken by balance *A*. The readings of *B* and *C* give the magnitudes of the components, and the reading of *A* the magnitude of a force which is equal and opposite to the resultant (as in Ex. 3), and whose point of application is the same as that of the resultant. With these data we can investigate the relations, as regards magnitude, point of application, etc., of parallel components and resultant.

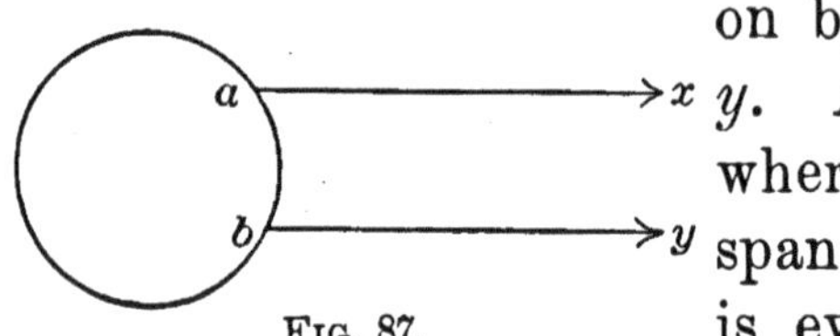

FIG. 87.

EXPERIMENT 1.

Apparatus.—Three 24-lb. balances; meter-stick; fish-line; means of suspending meter-stick (the whole arranged as in Fig. 90).

OBJECT.—When two parallel forces act on a rigid body, to determine (1) the relation of the resultant to the components in magnitude and direction; (2) the position of the point of application of the resultant with reference to those of the components; (3) under what conditions equilibrium may be obtained; and (4) under what conditions the body tends to rotate.

MANIPULATION.—The readings of *B* and *C*, Fig. 88, when forces are applied will not represent the true forces, because the balances themselves weigh something, and alone

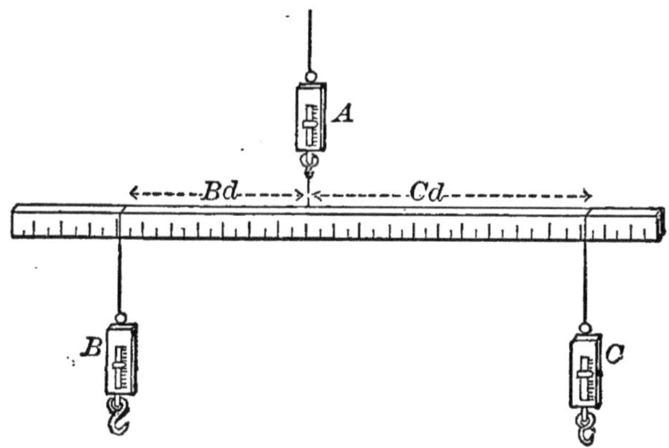

FIG. 88.

exert a slight downward pull. To correct for this error, detach the apparatus from the balance *A*, slip the loops of string off from the meter-stick, and weigh balances *B* and *C* on *A*. Take half the weight so obtained as the weight of one balance.* Again, since balance *A* has to support the weight of the meter-stick in addition to the forces used, the weight of the meter-stick must be found. Remove balances *B* and *C* from *A*, suspend the meter-stick alone, and record the reading of *A*. Since this reading also gives *A*'s increased reading due to the zero, error, we have the total amount to be subtracted from each subsequent reading in order to get the true force exerted on *A* by the combined pulls of *B* and *C*. When ready for work, the table of errors should stand something like this:

Correction for weight, *B*....*C*....*A*....
" to zero, *B*....*C*....

Record on which side of the pointer the readings were taken.

Slide the loops of the strings from *B* and *C* to such

---

* These weighings could be still better performed with a 64-oz. balance.

points on the stick that the distance of each loop from the centre is over 10 cm. Pull down steadily on the two lower balances, being careful to pull vertically, until each balance reads more than five lbs. Read the three balances, and note the direction of balance *A*. Place the loops at different points on the stick, and try again. Try a case where the distance on one side is twice that on the other. Try a case where the distances are both alike. Try other cases, making six in all. Great care must be used in reading the balances as accurately as possible, and always from the side of the pointer used in getting the readings for errors. Tabulate the results as follows:

TABLE I.

Read. *A*.	Read. *B*.	Dist. *B*.	Read. *C*.	Dist. *C*.	Direc. *A*.	Trial.

While you have equilibrium, increase the pull of one balance. Note what happens. Decrease, and note again.

CALCULATION.—The balance-readings have three errors for which corrections must be made: (1) Owing to use, the balances all read high; (2) *A* reads high because of the weight of the apparatus; (3) The readings of *B* and *C* show the forces applied less the weight of the balances themselves and their zero-errors. Correct *A* by subtracting total weight-error. Correct *B* and *C* for zero-error, and add their-weight. Record the results as follows, giving corrected readings:

TABLE II.

Read. *A*.	Read. *B*.	Read. *C*.	Dist. *B*.	Dist. *C*.	$C \times Dc$	$B \times Db$

The values obtained by multiplying the magnitudes of parallel forces by their "leverage," that is, the distance of the point of application of each force from the same point of reference, as $C \times Cd$, Fig. 88, are called the *moments* of the forces.

QUESTIONS.—1. Can you see any relation between *A*, *B*, and *C* as regards (*a*) direction; (*b*) magnitude? 2. Is anything noticeable on comparing the moments of *B* and *C* for each case? 3. Can you make out any relation between *B* and *Db* that also exists between *C* and *Dc*? 4. When the stick is at rest, how do the moments of *B* and *C* compare? 5. When the moment of one force was made larger than the other, by suddenly pulling down on one balance, what happened? 6. Can you name a case in which the principle of this experiment is used?

EXERCISE 5.

THE INCLINED PLANE.

Preliminary.—We know that machines enable us to make a small force overcome a greater one, but do they save work? When a force is overcome by the aid of a machine, is any less work done than when the force is overcome directly? For example, if a weight is lifted directly 10 ft., or lifted the same distance by a machine, is the work done in the former case any more than that done in the latter? To answer this question, we may raise a known weight a known vertical distance by means of a machine,—say an inclined plane,—and compare the work done in this case with that done in lifting the body vertically the same distance. For this purpose the apparatus illustrated in Fig. 89 may be used. The board *B* forms the inclined plane, and may be set at any angle by adjusting the rod *R* of the support *S*. The weight to be raised is the loaded carriage *C*, and the force required to pull it up the plane may be obtained from the reading of the balance *B*. We can get the work done in raising the body vertically by multiplying the total weight of the body by the distance it is raised (say *ab* in the figure), and the work done in raising it by the plane by multiplying the force required to pull

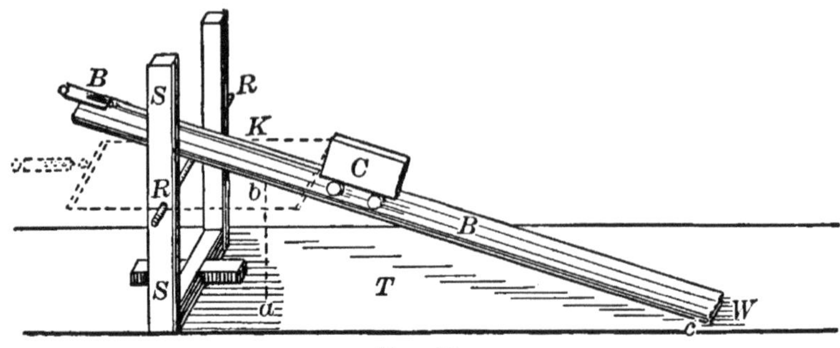

FIG. 89.

the body up the plane by the distance that it must travel on the plane, *cb* in the figure, to reach the required height.

EXPERIMENT.

Apparatus.—Board of Exercise 2; support; weighted carriage; 24-lb. balance; cross-stick and cord; meter-stick; T-square or plumb-line.

OBJECT.—1. To study the laws of the inclined plane. 2. To see if a machine, taking the inclined plane as an example, saves work.

MANIPULATION.—Get the weight of the carriage and load together.* In case no suitable scales can be had, the weight may be found with sufficient accuracy by means of the spring-balance. Adjust the support so that the board *B* is inclined at an angle of about 30 degrees. Hook the balance into the loop at the end of the cord connected with the carriage. Let one student hold the balance firmly in both hands, and draw the carriage up the plane at a uniform rate of speed. The back of one hand should rest on the board upon which it slides, and his *entire* attention should be devoted to holding the balance firmly, and pulling the carriage *at a uniform rate.* He should pull exactly parallel to the plane, and hold the balance so that it will not bind, *keeping his eyes away from the balance altogether.*

* This weight may be determined once for all, and marked upon the carriage.

Meantime let the second student keep his eye directly over the balance and take as many readings as possible. Owing to little inequalities in the board, etc., these readings will vary slightly, and the average position of the index must be determined as accurately as possible. A number of trials should be made, the students alternating in reading the balance, etc., and different parts of the board being used. The average of these trials, after the following corrections have been made, will give very closely the force required to pull the carriage and load up the plane.

The force used to pull the body up the plane is not the true force required, because a portion of it is expended in overcoming the friction of the carriage-wheels, etc. To correct for this slight excess in the reading of the balance, let the body slide down the plane at the same rate of speed at which it was pulled up, reading the balance as before. When the necessary correction for the zero error of the balance has been made, the average of the two readings will be very nearly the true force required.

Carefully measure the distance the body must travel on the board for a vertical rise of one foot, *ab* in the figure, or any vertical distance suitable to the apparatus. If a "T-square" can be obtained, the vertical rise taken may be marked off upon one side of it, starting from the outer edge of the cross-piece. The T-square is then placed vertically upon the table resting against the board, as shown in Fig. 90, and slid along until the mark indicating the vertical distance coincides with the lower edge of the board. The distance from this point to the lower corner of the board is the length of the plane to be measured. If no T-square is available, a plumb-line of the required length may be

FIG. 90.

until it just touches the table. The method with the T-square is the best.

Change the angle of the board and repeat the experiment. Tabulate results as follows:

TABLE I.

Approx. Angle of Board.	Up-force.	Down-force.	Vertical Rise of Body.	Distance on Plane.

CALCULATION.—From the results of Table I, calculate and arrange the data under the following headings, where $W =$ the weight of the body, $F =$ the true force required to pull the body up the plane, $Dv =$ the distance the body was raised vertically, and $Dp =$ the distance the body moved on the plane for this vertical rise.

TABLE II.

W.	*F.*	*Dv.*	*Dp.*

QUESTIONS.—1. What effect has the angle of the board on the force required to pull the body up the plane? 2. What effect has the angle of the board on the distance the body must travel on the plane for a given vertical rise? 3. Can you make out any relation between W and F that also exists between Dv and Dp? 4. Calculate for each case the work done in raising the body the given vertical distance, say 1 ft., and the work done in raising it the same distance by means of the plane.* 5. Can any rela-

* In the first case, this equals

$$W \times Dv,$$

and in the other

$$F \times Dp.$$

These may be calculated in any units of work, but the same units

tion be made out between the work done in raising a body a certain distance vertically and the work done in raising it the same distance by the plane? Does this hold for more than one angle of the plane? 6. Does the machine save work? (i.e., is less work done when the plane is used than when the body is lifted vertically?) 7. How does the machine help you?

EXERCISE 6.

THE WEDGE AND THE SCREW.

Preliminary.—In the following exercise it is desired to discover whether the facts just observed also hold when the force used to draw the body up the plane is applied parallel to its base, as in the wedge and the screw, instead of to its length, as in the inclined plane. For this purpose the same apparatus may be used, the force being applied by pulling the balance parallel to the table instead of to the board.

EXPERIMENT.

OBJECT.—1. To study the laws of the wedge and screw. 2. To see if the inferences drawn for the inclined plane will hold for the wedge and screw.

MANIPULATION.—Attached to the front of the carriage, *C*, Fig. 89, is a stick long enough to extend beyond the support at both sides. A second stick of the same length is connected with the first by cords, as shown by dotted lines, and the balance is attached to the centre of the second stick. As the connecting cords pass outside of the support, the body can be pulled up and down the plane without difficulty, by a force parallel to the table. Let one student hold the balance in both hands, face up, and pull the carriage up the plane, keeping the strings parallel to the table by raising his hands as the body rises on the plane. A second student should stand a short distance away, where he can see if the pull is actually parallel to the table. The students should alternate in reading the balance.

Make the observations, records and corrections, and answer the questions, as in Ex. 5.

EXERCISE 7.

LAWS OF THE PENDULUM.

Preliminary.—The apparatus used in this exercise is shown in Fig. 91. Near the edge of the support *S* (which may be the edge of a table or shelf) is screwed a spool *S′*. The screw is "set up" until the spool turns with considerable friction. A string is wound around the spool and is held in place by passing through the slot of a screw, *R*, inserted horizontally in the edge of the support. The lower end of the string passes through a hole in a metallic ball *B* which forms the pendulum-bob. The length of the pendulum may be varied by turning the spool so as to wind or unwind the string. Small adjustments are best made by gently turning the spool.

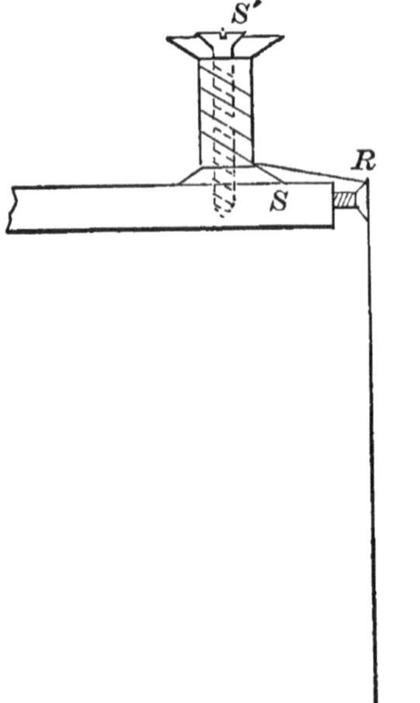

FIG. 91.

EXPERIMENT.

Apparatus.—Two iron balls of different sizes, or iron and wooden balls; apparatus for suspension (screw, spool, fish-line); callipers or rectangular blocks to get diameters; meter-stick; time-piece.

OBJECT.—To determine the effect on the number of vibrations of a pendulum, of (1) length of arc, (2) length of pendulum, (3) weight of bob.

MANIPULATION.—*Part* 1. Length of arc. Make the length of the pendulum about 50 cm. Count the number of vibrations it makes in two minutes when swinging through an arc of not over 30 cm. Increase the arc to, say, 60 cm., the length of the swing being estimated by the eye. Determine as before the number of vibrations in two minutes. Compare the number of vibrations a minute in each case and draw your inference. Record the results as follows:

TABLE I.

Arc.	No. Vibrations in 2 min.	No. Vibrations per min.
Long		
Short		

From the observation of the data, infer the effect of length of arc.

Part 2. Length of pendulum. Observe the general effect of changing the length of the pendulum. Record your observations and inference in general terms.

To make the quantitative determination, we must compare the number of vibrations in equal times of pendulums of different measured lengths. The length of the pendulum is the distance from its *centre of gravity* to its point of support. The string having practically no weight, the centre of gravity of the pendulum corresponds with that of the ball, and is in the centre of the ball. To find the length of the pendulum, therefore, we must measure the distance along the string from the lower edge of the screw to the top of the ball, and add to it half the diameter of the ball. Place the lower end of the meter-stick on top of the ball, and bring its graduated edge up to the string. Measure the distance to the edge of the screw three times, reading to millimeters. The average of these three readings will probably be nearly correct. Get the diameter of the ball, where the hole is, by placing it between the jaws of the callipers and then applying the callipers to the scale. Repeat three times and average.* The length of the string plus half the diameter of the ball is the length of the pendulum.

Set the pendulum swinging through an arc of about 2 ft. Measure the number of vibrations in 1, 2, and 3 minutes. Reduce the length of the pendulum about half, measure exactly, and repeat the counting. Record results as follows:

TABLE II.

Length of String.	Av. L. of String.	Diam. of Ball.	Av. D.	$\frac{D}{2}$.	True Length.	No. Vib.	Time.	Av. No. Vib.

Part III. Weight of the bob. Substitute for the ball on the pendulum a larger ball. Carefully determine the diameter of the second ball, as before. Make the length of the string such that the length of the pendulum is the same as before, i.e., the length of the string plus the radius of the ball equals one of the lengths used in Part II. Determine the number of vibrations in 1, 2, and 3 minutes as before. Record as in Part II. Compare the average number of vibrations per minute with that of similar length in Part II.

CALCULATION.—Divide the smaller length in Part II by the larger. Divide the smaller number of vibrations by the larger. Compare the ratios so obtained. If they do not agree, try squaring or cubing one of the ratios, and find the values approximating most closely. Infer the law, and express it as a formula.

---

* Or use the method given in notes on Mensuration.

EXERCISE 8.

ACTION AND REACTION.

Preliminary.—In this exercise we experiment with two bodies by giving them various amounts of energy and allowing them to collide. The special point of the experiment is to compare (1) the energy possessed by each body before and after the collision, (2) the energy lost by one with that gained by the other, and (3) the results of trying elastic and non-elastic bodies. The apparatus used is shown

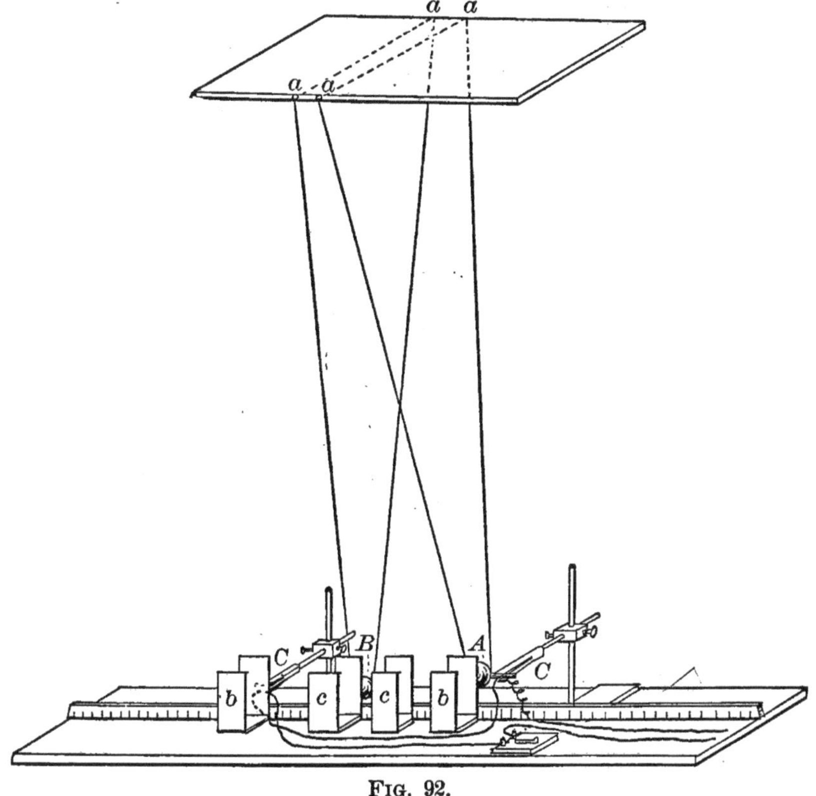

FIG. 92.

in Fig. 92. Two ivory balls, *A* and *B*, are so suspended that they can swing in one plane only, and will collide at the lowest point of their arc. They may be drawn aside any desired distance and held there by the electro-magnets *CC*, and released by breaking the circuit. The distances they are drawn aside may be read on the scale by aid of the cards *bb*. The momentum in each case is taken as the product of the weight of the ball multiplied by the distance it moved.

EXPERIMENT 1.

Apparatus.—Two ivory balls; No. 30 wire; electro-magnets; board; two meter-sticks; electric current; circuit-breaker; tacks; made into apparatus as shown; scales and weights (if balls are not of known weight).

OBJECT.—To compare (1) the algebraic sums of the energies possessed by two bodies before and after collision, (2) the results of trying elastic and non-elastic bodies, (3) the direction of the action and reaction, (4) the energy lost by one with that gained by the other.

MANIPULATION.—A little adjustment of the magnets as regards their position in the plane of oscillation will enable the student to release the balls so that they will strike squarely upon each other. By means of the index-cards *cc*, determine on the meter-stick the position of the balls when at rest. One ball being at rest, place the magnet so as to hold the other ball about 10 or 15 cm. aside. Read the position of this ball by the index-card *b*. Break the circuit, thus allowing one ball to strike the other, and notice the distance traversed by each ball after collision. Place an index-card at each of these points, and repeat the experiment. Sighting across each index, note the exact position reached by the centre of the corresponding ball, and place the index at this point. Try again until each index marks the exact position reached by centre of the corresponding ball after collision. As the magnet remains in the same place all the time, the distance traversed by the ball before collision is always the same, and need not be measured again. Following this method, try the following cases, calling the small ball *B*, and the large ball *A*:

Case 1. *A* in motion, *B* at rest. Try two cases, and repeat with *B* in motion and *A* at rest.

Case 2. Release both balls at once. Try covering one ball with a thin layer of putty. The weights of the balls should be determined in each case.

Compare the momenta before and after collision in each case. Distinguish between the behavior of elastic and inelastic substances under these circumstances. In calculating the momenta, call the movement towards the right hand +, towards the left hand —. Tabulate results as follows:

Weight of *A*.... Weight of *B*....

No. Trial.	Pos. of *A* at rest.	Pos. of *A* drawn aside.	Dist. trav. by *A* before coll.	Pos. of *B* at rest.	Pos. of *B* drawn aside.	Dist. trav. by *B* before coll.	Pos. reached by *A* after coll.	Dist. trav. by *A* after coll.	Pos. reached by *B* after coll.	Dist. trav. by *B* after coll.

From the above data calculate the momenta before and after collision in each case, being careful to keep the proper signs. Arrange results as follows:

Case 1. Momentum of *A* before collision =
" " *A* after " =
" " *B* before " =
" " *B* after " =
Algebraic sum =

From these data answer the questions given in the object.

SUBSTITUTE EXPERIMENT.

Apparatus.—Two pint tin pails, suspended and adjusted as shown; heavy bodies to load pails; cotton string; candle; two meter-sticks; coarse scales.

OBJECT.—(1) To compare the values of action and reaction when two bodies are acted upon by the same force, and (2) to study the effect of varying the weight of the bodies.

MANIPULATION.—Set up the apparatus as shown in Fig. 93. Call the pail with the spring *A*, and the other *B*. Detach *B* from the wires, and make it weigh 250 grams (cover *on*). Compress the spring and tie it with a string.*

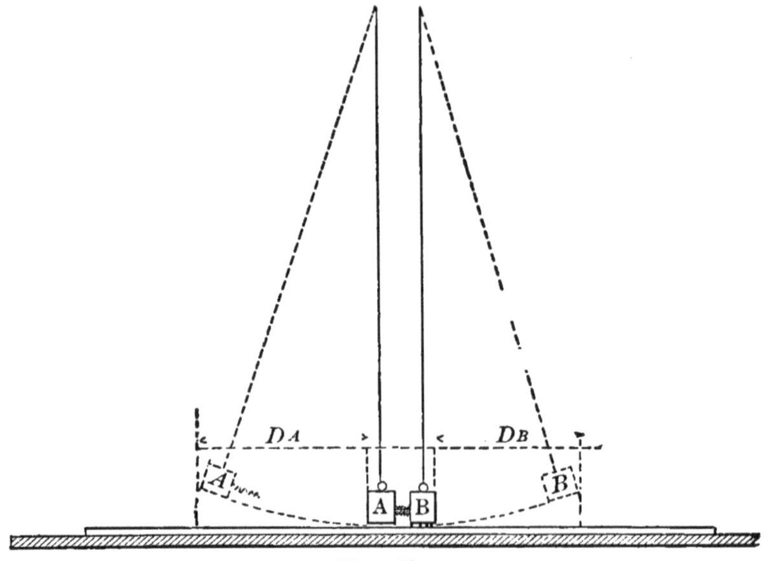

FIG. 93.

In the same way make *A* weigh 500 grams. Attach both the pails to the wires, being sure that they hang vertically. Holding the eye directly above the outer edge of each pail in turn, read its position on the meter-stick underneath, and record. Now place a meter-stick under the edge of each pail, and burn the string. As the pails swing out keep a pencil directly under the outer edge of each one, and stop them at the extreme point of the swing. While the pails return, hold the pencils steady at these points, read their positions on the meter-sticks, and record. Repeat three times; then make pail *B* 500 grm., and try again, say three times more. If possible, try again with 750 grm. Tabulate results as follows:

* It is a good plan to hold the hand between the pails when they come together, so that they will not strike each other with a heavy blow.

TABLE I.

Wt. *A*.	Wt. *B*.	Read. *A* at rest.	Read. *B* at rest.	Read. *A* at end of swing.	Read. *B* at end of swing.

From the data in Table I calculate the following table:

TABLE II.

Wt. *A*.	Wt. *B*.	Dist. *A* moved.	Dist. *B* moved.	Wt. *A* × Dist. *A*.	Wt. *B* × Dist. *B*.

QUESTIONS.—1. Can any relation be made out between the weights of the pails and the distances that they moved? 2. What relations appear between the numbers in the last two columns in Table II? 3. Calling w and W the weights of the two pails, and d and D the distances moved, can the results be expressed as a formula? 4. Can any law be inferred for the relation of action and reaction regarding (*a*) direction, (*b*) extent of motion?

EXERCISE 9.

THE FORCE OF TENACITY.

EXPERIMENT.

Apparatus.—Wire of two sizes and two materials; half-spool screwed to table; 24-lb. spring-balance; meter-stick.

OBJECT.—To study the effect of length, cross-section, and material on the force of tenacity.

MANIPULATION.—*Part I.* Effect of Length. Pass one end of the wire twice around the half-spool with a "round turn," making it lie close to the spool all the way; then twist the end of the wire around its main part, as indicated in Fig. 94. The spool is screwed to the table.

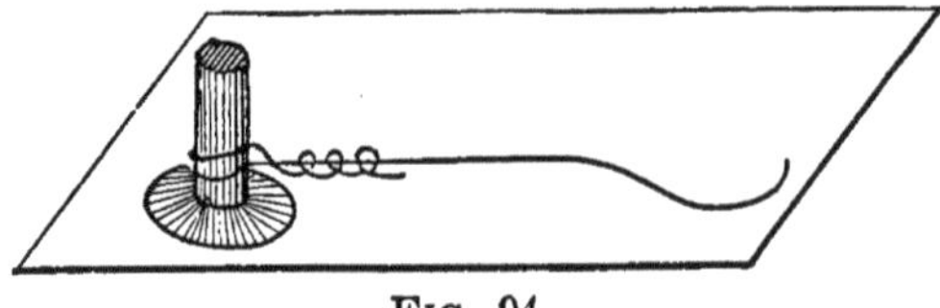

FIG. 94.

Cut off the wire about one meter from the spool, and attach the end to the balance-hook, precisely as shown in Fig. 94. Examine the wire to make sure that it is free from "kinks," and measure its length. Then, holding the balance horizontally in the palm, move the hand along with its back on the table so as to produce a *steady* pull on the wire. Steadily increase the pull until the wire breaks, meantime watching the index of the balance *all the time* so as to know where it stands when the wire breaks. The first trial will tell about what part of the scale to watch in subsequent trials. Make four trials besides the first, and try to read to $\frac{1}{4}$ lb., especially in the third and fourth trials. Take a different length for each trial, and record the length of each piece.

Part II. Effect of Cross-section. In the same manner determine the tenacity of the second size of wire, making four trials.

Part III. Effect of Material. Determine in the same way the tenacity of the wires of different material, making four trials.

Tabulate results as follows:

No. Trial.	Length of Wire.	Cross-section.	Bk.Wt. on Bal.	Correct Bk.Wt.

The values in the last column are the balance-readings corrected for zero and position errors.

QUESTIONS.—1. Can you make out any relation between length and tenacity? 2. Between cross-section and tenacity? 3. Has material any effect? 4. What cross-section of one wire would give the same tenacity as another wire of different cross-section and material?

EXERCISE 10.

THE FORCE OF ELASTICITY.

Preliminary.—A body whose shape has been changed by the action of a force is said to be *distorted*. A body which tends to take its former shape when distorted is said to be *elastic*. The force with which a distorted elastic body tends to resume its original shape is called *the force of elasticity*. A body which may be considerably distorted with very little force is said to be very elastic, or to have *low elasticity*. Rubber is an example of low elasticity. A body which gives considerable force of elasticity on small distortion is said to have *high elasticity*. Steel is an example of high elasticity. The following exercise investigates the effect of (1) degree of distortion, (2) cross-section, and (3) length on the magnitude of the force of elasticity in a solid. For this purpose the apparatus in Fig. 95 is used. The rubber strip *S* is fastened at its upper end, and may be stretched (distorted) by weights placed in the scale-pan *E*, which is attached to its lower end. Since, after the strip has stretched and is at rest, the *up* pull due to the elasticity of the rubber, and the *down* pull due to the weights, must be equal, the weights measure the force of elasticity. The total amount of distortion is measured by reading the position of the point *a* on the meter-stick, by means of the reading-card *C*. The distortion of two lengths, one twice the other, may be compared by reading first from the point *a*, and then from point *b*. The position of

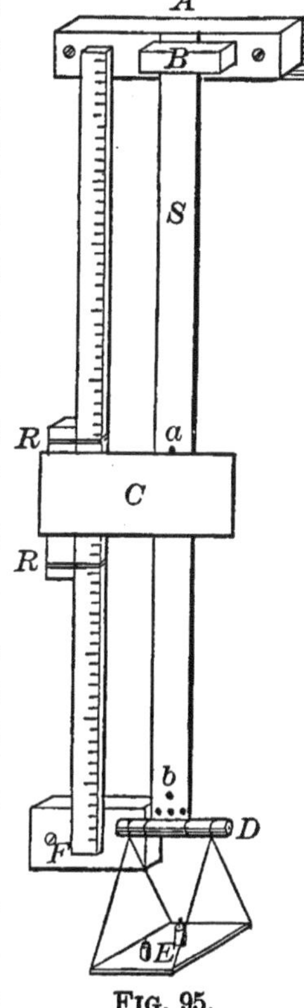

FIG. 95.

the card *C* may be changed by slipping off the rubber bands *RR*, and sliding it along the meter-stick.

EXPERIMENT.

Apparatus.—Apparatus for the experiment as shown; weights 10 to 200 g.; screw-driver.

OBJECT.—(1) To see if any relation can be found between the amounts of distortion of an elastic body and the corresponding magnitudes of the force of elasticity. (2) To observe the effect (*a*) of cross-section, (*b*) of length.

MANIPULATION.—*Part I.* Adjust *C* so that the upper edge of the card touches the lower edge of the mark *a* on the rubber strip. Hold it there by hand or by the rubber bands, and read and record the position of the upper edge of the card on the meter-stick. Gently place 20 gr. on the scale-pan, and after the strip has come to rest, again adjust *C*, and read the position of *a* and record. Repeat with 40 gr. in the pan, and so proceed for ten readings in all (up to 200 gr.). Record as follows:

Weight used.	Position of *a*.	Distortion of *S*.*

Part II. Repeat Part I, reading from *b*, and record in the same way.

Part III. Detach the strip used in Part II, and substitute the wide one. Repeat with this strip. When read from the mark upon it, this gives the same length and cross-section as the one in Parts I and II.

QUESTIONS.—What effect have length and cross-section? 2. Are all bodies equally elastic? (Answer from your gen-

* The osition of *a* for each wei ht minus that for no wei ht.

eral knowledge.) 3. What conditions have you found to affect the magnitude of the force of elasticity? Plot a cure from the data of Part I.

EXERCISE 11.

BOYLE'S LAW.

Preliminary.—We know that when pressure is exerted on a confined volume of gas its volume becomes less, and that when the pressure is diminished the volume increases. In the following exercise we wish to see if we can find any relation between different pressures and the corresponding volumes.*

The apparatus used is shown in Fig. 96. A glass tube, *ab*, is bent as shown. The short arm is closed at the top, the long arm open. The gas to be experimented upon is confined at *b* in the short arm by pouring mercury into the long arm. The pressures can be changed by using different depths of mercury, the volume of gas can be read on the scale *d*, and the depths of mercury on the scales *d* and *e*. The two cards *ff* help in reading the levels. As we know from the laws of liquid pressure, the depth of mercury causing the pressure will be the *difference* in the heights of the two columns; and since the atmospheric pressure is exerted on the top of the column in *a*, the total pressures will equal the

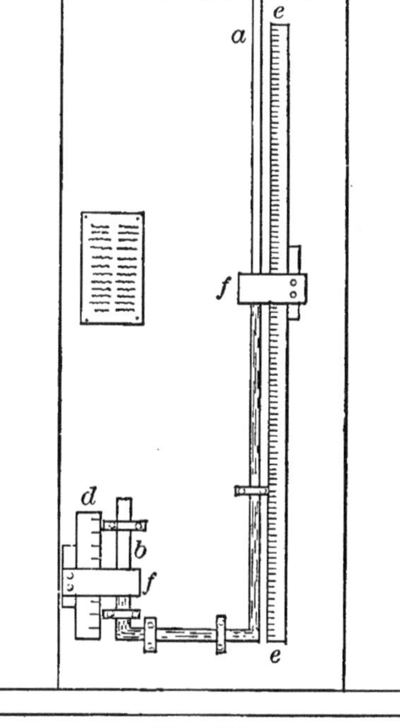

FIG. 96.

* Suggest a method for such an experiment.

height of mercury, causing the pressure plus the height of the barometer, whose reading must be known. Since changes of temperature affect the volume, the temperature must be practically uniform during the exercise.

EXPERIMENT.

Apparatus.—As shown (tube; support; scales; reading-cards); about 500 gr. mercury (clean and dry); barometer; feather and rod to remove air.

OBJECT.—To discover some relation between the volume of a confined body of gas and the pressure exerted upon it.

MANIPULATION.—Read the barometer. (See the card of instructions tacked over the instrument, with instructions for reading vernier, etc.) The barometer reads in inches; as the other readings will be in centimeters, convert the barometer-reading to centimeters by the following formula:

$$\frac{\text{Bar.-read. in inches}}{.3937} = \text{Bar.-read. in cm.}$$

Read the thermometer. Place the glass funnel (be sure that it is clean and dry) in the open end of the long arm, and carefully pour in mercury until the bend is covered and the mercury stands two or three cm. higher in the long arm than in the short arm. Tip to the left the apparatus as it stands in Fig. 96, and allow some air to escape from the short arm; place the apparatus upright again, note the levels, and, if necessary, repeat until the mercury stands at about the same level in both arms. If the level of the mercury in the short arm is still below the upper edge of the horizontal part of the tube, add a little more mercury. If too much air has been allowed to escape, tip the apparatus to the right and let some air run in. Working in this way, get the level of the mercury in the short arm above the curve, and the mercury in the long arm from 1 to 3 cm. higher. Then read the position of the top of the meniscus of each mercury-column on its scale. Place the funnel again in the top of the tube, and carefully pour in mercury until the level of the column in the long arm has been raised about 15 cm. Remove the air-bubbles and again read the heights in both arms on the scales. With the same precautions add another 15 to 20 cm. height to the long-arm column. Again read the levels. So proceed till the *difference* in the levels of the columns in the two arms is about 75 to 85 cm. Again read the barometer, and if you find any difference in the barometer-readings before and after these operations, use the average of the two. Read the thermometer, and if you find a small change in temperature, average the readings; if a large change, report the fact at once. During the progress of the experiment, watch the thermometer and note any change of temperature. Now you have the following data:

(*a*) Barometer-reading before and after.
(*b*) Thermometer-reading before and after
(*c*) Levels of mercury in the short arm.
(*d*) " " " " " long "

Arrange (*c*) and (*d*) in a table as follows:

TABLE I.

Height in Short Arm.	Height in Long Arm.	No. Trial.

CALCULATION.—Arrange the results of the calculation in a table of five columns, as follows:

TABLE II.

Volumes.	Pressure.	Whole Press.	Ratios of Vol.	Ratios of Press.

The figures in column 1 are obtained from the card attached to the apparatus, the volumes in cubic centimeter corresponding to each reading on the short-arm scale, The higher the number on the scale the less the volume of air represented. The figures in column 2 are obtained by subtracting the readings of the short arm from those of the long arm. The figures in column 3 are obtained by adding to each number in the second column the height of the barometer in centimeters. This gives the total pressure of mercury in centimeters. The figures in column 4 are obtained by dividing each volume in turn by the smallest volume. The figures in column 5 are obtained by dividing each pressure (column 3) in turn by the smallest pressure. Carry out both these ratios to the second place of decimals.

QUESTIONS.—1. Is there any law as regards the ratio of the volume to the pressure of a gas, the temperature being constant? 2. If so, what is it? 3. Why read the barometer twice? 4. Why read the thermometer twice? 5. What keeps the mercury from completely filling the small arm? 6. Is it of the nature of a push? 7. Is it a force? 8. What is the name of that force? 9. In this experiment what is the *stress*? 10. In this experiment what is the *strain*? 11. What sort of elasticity has a gas? 12. What principles, learned in previous experiments, have you made use of in this one? Plot a curve from the data.

EXERCISE 12.

SPECIFIC GRAVITY WITHOUT SCALES OR WEIGHTS.

EXPERIMENT 1.

Apparatus.—Suspended meter-stick; two stones or other bodies of about equal weight; cords; vessel of water.

OBJECT.—To determine the specific gravity of a solid by the principle of moments.

MANIPULATION.—Call the body whose specific gravity is to be determined *A*, and the other *B*. Suspend *A* about 25 cm. from end of the meter-stick, Fig. 97, having the loop of the suspending string tight enough not to slip during the experiment, Slide the weight along until it balances *A*, and note its distance from the support *S*. Bring a vessel of water under *A*, and when *A* is completely submerged, and clear of the sides and bottom of the vessel, move *B* until it balances it again, and note its distance from *S*. You have now the data for determining the specific gravity of *A*.

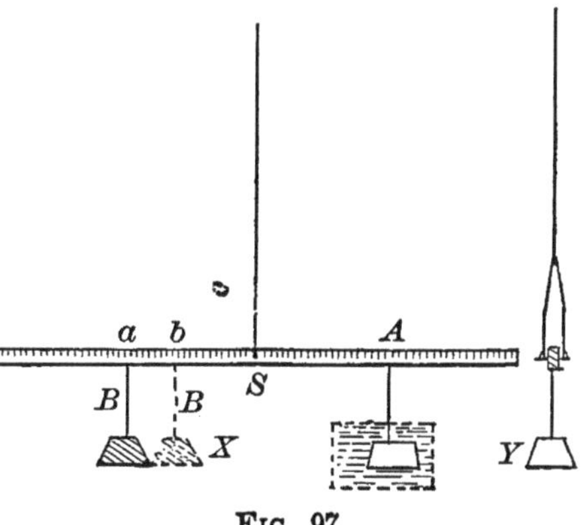

FIG. 97.

CALCULATION.—In Fig. 97 aS represents the weight in air, bS represents the weight in water; so $aS - bS$ or ab represents the loss of weight, and specific gravity $= \frac{aS}{ab}$.

EXPERIMENT 2.

Apparatus.—Long spiral spring; meter-stick; clamp; body whose specific gravity is to be determined; vessel of water; string or wire.

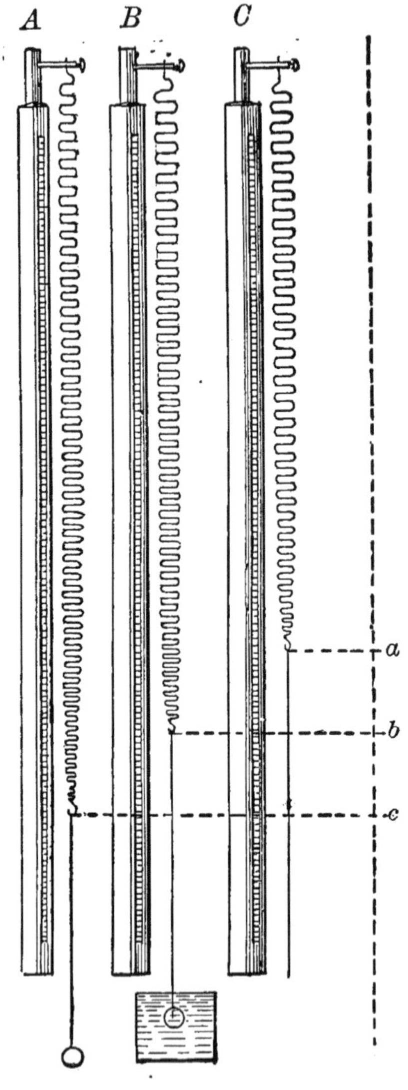

OBJECT.—To determine the specific gravity of a solid body.

MANIPULATION. — Note length of spring, with suspended wire only attached, Fig. 98, *C*. Attach the body to the lower end, as in Fig. 98, *A*, and record the length of spring. Bring the vessel of water under the body, immerse the body with the usual precautions, and again record the length of the spring *B*.

CALCULATIONS.—Reading of reference-marks:

Spring alone =
With body attached =
Body immersed =

ac represents wt. in air.
ab " " " water, and
$ac - ab$ represents loss of wt. in water, and
Specific gravity =

$$\frac{ac}{ac - bc} = \frac{ac}{bc}.$$

FIG. 98.

In same way find specific gravity of some liquid.

EXERCISE 1.

FOCI OF LENSES.

Preliminary.—When the light from a luminous body passes through a double convex lens, the point at which all the light is concentrated is called the *focus* of the lens. The distance from the centre of the lens to the focus is called the *focal length*. The following exercise investigates the effect on the focal length of the distance of the object from the lens. The apparatus shown in Fig. 99 is used.

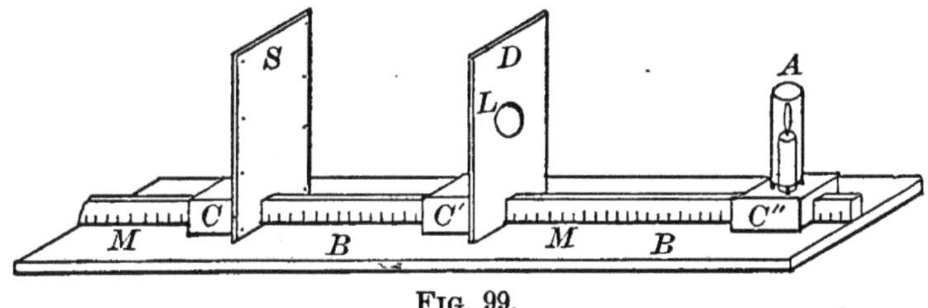

FIG. 99.

For the luminous object near the lens we use a light *A*, represented as a candle mounted upon a block *C''*, which slides upon the meter-stick *M*. The lens *L* and the screen *S* are similarly arranged. The position of the lens can be so adjusted that the image of *A* falls on *S* when *A* is at different distances from the lens, and the distances *AL* and *LS* can be read on the meter-stick.

EXPERIMENT.

Apparatus.—As in Fig. 99, so arranged that an image of an object at a considerable distance may be obtained.

OBJECT.—To study the effect of the distance of the object from the lens on its focal length.

MANIPULATION.—Bring the outer ends of blocks *C* and *C''* even with the ends of the meter-stick. Light *A*, place it in the centre of block *C''*, and move the lens to a point about 20 cm. from *A*. Slowly slide the block *C'* towards *S* until a position is found which gives a sharp image of *A* on the screen. Record the readings of the right-hand sides of *S* and *D* and of block *C''* (in this case, 0). Move the lens towards *S* until the image is lost, then slowly move it back towards *A* until a sharp image is again obtained, and take the reading of D as before. Gently slide block *C''* 10 cm. nearer the lens, and repeat. So continue at intervals of 10 cm. until an image can no longer be obtained. To test the case when the object is as far from the lens as possible, extinguish the light and remove it, together with block *C''*. Arrange the rest of the apparatus so that the light from some distant object* can pass through the lens and fall upon the screen. Find as above the position of D that gives a sharp image. Record as follows:

TABLE I.

Read. of *S*.	Read. of *D*.			Read. of *A*.
	R.	L.	Av.	

In the first and third columns place the readings of right-hand sides of *C''* and *S*, respectively. In the second column *R* and *L* are the readings of the right-hand side of D when moved from the right or from the left to the position giving a sharp image.

QUESTIONS.—1. Is the distance of the lens from the

* The distance of this object, which should be over 30 meters from the lens, should be recorded, whether measured or estimated by the eye.

screen the same whatever the distance of the luminous object from the lens? 2. What other inferences can you draw from the results of the experiment?

When the object is so distant that the rays of light from it are practically parallel where they strike the lens, as in the case of the distant object in the experiment, the distance from the centre of the lens to the image is called the *true focal length* of the lens. When the object is so near the lens that the rays of light are not parallel, this distance is called the *conjugate focal length.*

CORRECTIONS.—In order to make any exact comparisons, it is necessary that the true positions of A and of the lens should be determined. To get the true position of A, add one half the length of block C'' to the readings of A in Table I. For the position of the centre of the lens add one half the width of the board **D** to the average readings for **D** in the same table. Record as follows:

TABLE II.

True pos. A.	True pos. L.	Pos. S.	Dist. AL.	Dist. LS.

CALCULATION.—Taking the first case, express the fraction $\frac{1}{AL}$ as a decimal carried out to four significant figures. Express the fractions $\frac{1}{LS}$ and $\frac{1}{\text{Focal length}}$* in the same way. Do the same for the other cases. Tabulate as follows:

---

* As obtained in the trial with the distant object.

TABLE III.

$\frac{1}{AL}$	$\frac{1}{LS}$	Sum of the two.	$\frac{1}{F.L.}$

QUESTIONS.—1. From the study of Table III can you make out any relation between the distance of the object from the lens (AL), the conjugate focal length (LS), and the true focal length of the lens? If so, letting F = true focal length, f = conjugate focal length, and D = the disance of the object from the lens, express your inference as a formula.

EXERCISE 2.

DISTANCE AND INTENSITY OF LIGHT.

Preliminary.—By the *intensity of light* is meant not the brightness of the source of light itself, as a lamp-flame, white-hot carbon, etc., but the degree to which its light illuminates a given body. It is a well-known fact that the illuminating power of a light decreases as the distance of the illuminated body from it increases. The light from a large lamp and from a candle may be of equal intensity if the illuminated body be much nearer to the candle than to the lamp. The following exercise investigates the effect of distance on intensity of light, with a view to seeing if any fixed relation can be made out between them. In order to do this, we determine the distances at which lights of different known powers give light of the same intensity. For this purpose we use an instrument called a *Bunsen Photometer*, whose operation is based upon the fact that when a piece of paper having a paraffine spot in the centre is equally lighted on both sides, the spot can no longer be seen. The form used is shown in Fig. 100. The lights are placed upon the blocks *C* and *E* and the paper upon the block *B*. All the blocks slide upon the meter-stick. For lights of different intensities we use different numbers of candles, assuming that two candles give twice as powerful a light as one. Taking lights of different intensities, and finding the dis-

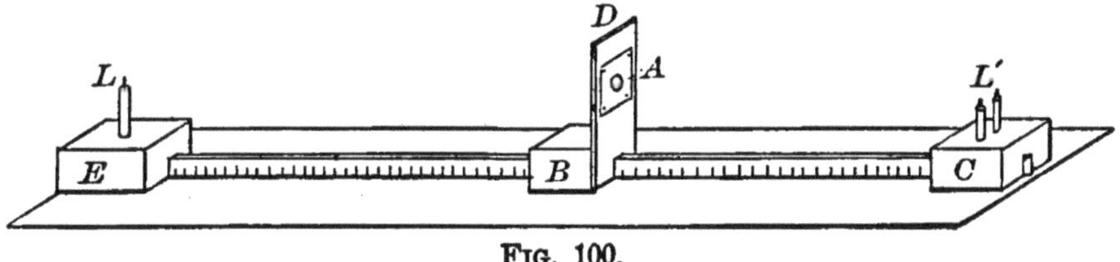

FIG. 100.

tance from each light to the paper when the paper has been placed so that the spot in its centre is no longer visible, we have the distances required for the different lights to give the same degree of intensity.

EXPERIMENT.

Apparatus.—Bunsen photometer, as in Fig. 100; 5 candles, 3 or 4 in. long; matches; scissors for trimming candle-wicks.

OBJECT.—To see what relation exists between the intensity of a light and its distance from the lighted object.

MANIPULATION.—Arrange blocks *C* and *E* at the ends of the meter-stick, as in the preceding exercise. Place one lighted candle on the block *C* and one on block *E*. Move the block *B* towards *C*, watching it from that side until the spot shows plainly; then slowly move it away until the spot just disappears. Have the eye about level with the spot, and the line of sight at an angle of about 30° with the meter-stick. If possible, shield the eye from the direct light of the candles by card-board screens. Read the position of the right-hand side of D. Next looking at *A* from the left-hand side, move *B* within 20 cm. of *E*; then back towards *C* until the spot just disappears. Record the reading of D. Move *B* within 20 cm. of *C*, and again determine the position where the spot is invisible when looked at from the right-hand side, as in the first case. Working in this way from each side alternately, make about ten readings in all. During the operation keep the candle-flames burning evenly, trimming the wicks if necessary.* In order that the effect of a variation in flame may be still more reduced, it is advisable to change the candles end for end of the apparatus, moving blocks and all. If this be done, take twelve readings in all. Repeat these readings first with two and then with four candles on *C*, setting the candles about 1 cm. apart in a line at right angles to the length of the meter-stick. Tabulate results as follows:

TABLE I.

Reading of *B*.			Pos. of Centre of L.	Pos. of Centre of L′.
R.	L.	Av.		

CALCULATION.—The readings for *B* are practically correct. The positions of the centres of the candles are found as in the preceding exercise. Record corrections as follows:

TABLE II.

Inten. of *L*.†	Inten. of *L′*.†	Dist. *LD*.	Dist. *L′D*.

The numbers in the last two columns are the distances from the centres of the candles to D.

QUESTIONS.—1. How would the distances compare if nine candles were used? Sixteen? Twenty-five? 2. Can you make out any uniform relation between the distances

* To protect the candles from draughts, lamp-chimneys may be placed over them, supported, of course, so that air may enter from the bottom.

† That is, number of candles used in each case.

and the intensities? If so, using I and I' for the intensities and D and D' for the distances, express your inference as a formula.

EXERCISE 3.

RADIATION OF LIGHT.

Preliminary.—It is generally known that light *radiates*, or spreads out in every direction, from a luminous point. The following exercise investigates the effect of distance from the light upon the degree of radiation. The light from a luminous body is allowed to pass through a small opening, and the degree to which it spreads is determined by letting it fall upon a screen whose distance from the light may be varied.

In order that the light should radiate from as nearly a point as possible, the light is placed in a box with a small hole opposite the flame on the side towards the screen.

EXPERIMENT.

Apparatus.—As in Fig. 99, the lens being removed and a piece of paper with a hole 2 cm. square tacked over the hole in D; a box fitted to hold the light; a short mm. scale; sharp pencil; some pieces of writing-paper about 3 × 3 in., one of which is pinned on the screen at the beginning of the exercise.

OBJECT.—To see if there is any uniform relation between the degree of radiation and the distance from the point of radiation.

MANIPULATION.—Light the candle and adjust the block supporting it so that the reading of the opening in the box through which the light passes is 5 or 10 cm. on the scale.* Place D 25 cm. and S 35 cm. from the hole. If the light is properly adjusted, a well-defined square of light will be thrown on the screen. Holding the screen firmly, place a ruler upon it, and with a sharp pencil rule a vertical line 2

* One method of doing this would be to hold the ruler firmly against the side of the box facing the screen, and adjust so that the right-hand edge of the ruler would come on the 5 or 10 cm. mark on the meter-stick.

or 3 cm. long on the left-hand edge of the square of light, using great care to get the lines exactly on the edge. Mark the right-hand edge in the same way. Set the screen 40 cm. from the hole, and rule two more lines. So proceed up to 60 cm. from the hole. Remove the square of paper pinned on the screen, pin on a fresh square and get another set of lines, working back from 60 to 35. Get four sets of lines in all, working back and forth twice. Measure carefully on all four pieces of pieces of paper the distance between the pairs of lines obtained for each position of the screen. Record as follows:

TABLE I.

Pos. of Point.	Pos. of Screen.	Dist. between Lines.				
		1st.	2d.	3d.	4th.	Av.

From these results calculate

TABLE II.

Dist. from Point to Screen.	Increase.	Av. Width of Square of Light.	Increase.

In the first column place the reading of the screen minus the reading of the left-hand side of the box. The figures in the second column are obtained by getting the differences between the different readings of the screen. In the third column place the average values obtained from Table I for the distance between the pairs of lines corresponding to each position of the screen, and in the fourth column obtain the differences in the same way as indicated for the second column.

QUESTIONS.—1. What inference can you draw regarding

the relations of distance and degree of radiation? 2. If D and D' be any two distances, and I and I' the corresponding intensities, can you combine them in a formula? Remember that for each distance the same amount of light will spread over a surface whose area would be proportional to the squares of the width of the square of light, and that the intensity of the light will become less in proportion to the surface illuminated.

EXERCISE 4.

CANDLE-POWER BY THE RUMFORD PHOTOMETER.

Preliminary.—The unit used in comparing the light-intensities is called *one candle-power,* written c. p. Candle-power is measured by determining the distances at which the source of light to be tested, and a standard candle specified by law, give lights of equal intensity. For example, if the light to be measured and the standard candle must be at the same distance in order to give light of equal intensity, the light of required intensity would be 1 c. p.

The form of photometer used in this exercise is shown

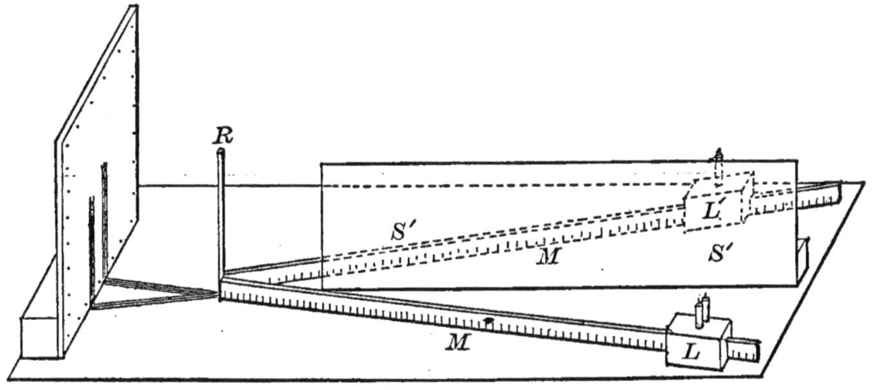

FIG. 101.

in Fig. 101, and is called the *Rumford Photometer.* Two meter-sticks *MM* carry two blocks *LL'*, one of which supports the standard reference-candle, the other the light to be measured, represented here by two candles. These blocks can be moved to any position on the meter-sticks. A card-board screen S' is placed between them, and an upright rod R at the left-hand ends of the meter-sticks. Each light will cast a shadow from this rod upon the screen at the left. When these shadows are equally black, both lights are of equal intensity at the screen.

EXPERIMENT.

Apparatus.—Rumford photometer as in Fig. 101; candle; light to be tested; scissors.

OBJECT.—To determine the candle-power of a light by the Rumford photometer.

MANIPULATION.—Place one candle on block L', light it, and set the block 30 cm. from the rod. Place on L the light to be tested. Allow the lights to burn for a few moments, and then move L towards the screen until the two shadows cast by the rod are equally dark. Read the position of the right-hand side of L. Move L towards the screen until its shadow is decidedly the darker (recollect that the shadows cross), then move it towards the right until the shadows are equal, and again record the reading of L. Move L' 10 cm. nearer R and repeat. Repeat again with L' 10 cm. still nearer. Record as follows:

Pos. of L'.	True Pos. of Candle.	Pos. of L.			True Pos. of Light.*	Dist. from Screen to* Candle.	Dist. from Screen to Light.
		R.	L.	Av.			

CALCULATION.—Using the formula of Ex. 2, all the values are given in the table above, except the intensity of the light to be tested, the intensity of the candle L' being taken as 1 c. p. Substitute these values for each case, taking the intensity of the light to be tested as X. Average the values so obtained, and put them down as candle-power.

* Distances of L and L' from rod, plus distance of rod from screen.

SOUND.

EXERCISE 1.

CONDITIONS AFFECTING PITCH.

Preliminary.—The following exercise investigates the conditions affecting the pitch of the note given by the vibrating wire. For this purpose the apparatus shown in

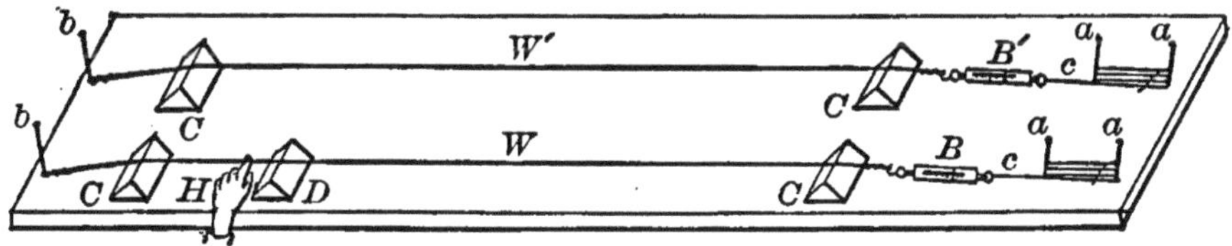

FIG. 102.

Fig. 102 is used. The wires *WW′* of different sizes are stretched by the balances *BB′*, the lengths used being those between the triangular blocks *CC′*. In order that the tension may be the same for different lengths, a fifth block, *D*, may be placed under the string, the wire being pressed down on *D* by the finger *H*, as shown in the figure. Cords *cc* wound around nails *aa* hold the balances at any desired tension.

EXPERIMENT.

Apparatus.—As in Fig. 102. Meter-stick.

OBJECT.—To observe the effect of (*a*) length, (*b*) tension, (*c*) size, on the pitch of the note given by a stretched wire.

MANIPULATION.—*Part I.* The apparatus being arranged as in Fig. 102, hold balance *B* in the left hand, take a turn of the cord *c* around the right-hand nail *a*, steadily pull to the left on *B*, and at the same time pulling the cord with the right hand until *B* reads 8 lbs. Remove the left hand, and let the whole strain come upon the cord. Make the cord fast by four or five crossed turns around both nails.

If the cord stretches so that the tension drops below 8 lbs., the operation must be repeated. In like manner put *W′* under the same tension. Bring D close to the left-hand block *C*; measure the length of wire between D and right-hand block by laying the meter-stick on the blocks parallel to the wire. Press the finger on the wire to the left of D, as at *H* in Fig. 102, and sound *W*. Shorten *W* by moving D to the right, and look for positions that give the first and second octaves of the note first obtained, measuring the length in each case. Move D back to its original position, and calling the note given by *W* "*do*" on the scale, look for positions giving the other notes in the gamut. If possible, carry these measurements through the second octave. Repeat, starting with a different length. Record as follows:

TABLE I.

Tension.	Length.	Note.

Part II. Remove D, sound *W*, and change the length of *W′* so that it gives the same note as *W*. Then find what tension on *one half-length* of *W* will give the same note as *W′*. The note given by *W′* is simply used as a reference; hence the length of *W′* need not be recorded. Shorten *W* 10 cm. by moving the block D, again set *B* at 8 lbs., and repeat the whole operation. Do this for several lengths, and record.

TABLE II.

1st Tension of *W*.	Tension of *W* for same Note on *half*-length.	Note that *whole*-length of *W* would give.*

* See Part I.

Part III. With a tension of 8 lbs. on each string, and with equal length of wire, compare the notes. Do this for several lengths and tensions. Record

TABLE III.

Length W.	Length W'.	Tension W and W'.	Pitch of note.

QUESTIONS.—1. Can you make out any relation between the length giving any note and that giving its octave? 2. Do the lengths giving the note on the scale bear any definite relations to the length giving the first note? If so, what? 3. Can you make out any connection between tension and pitch? 4. Between size and pitch? 5. Name some musical instruments in which these principles are used, and explain how they are applied.

EXERCISE 2.

VELOCITY OF SOUND.

Preliminary.—In the following exercise the velocity of sound is determined in two media—air and carbon dioxide. The method used is based upon the fact that if a tuning-fork be sounded at the mouth of a tube closed at one end, the length of the air-column which will reinforce the sound of the fork is one quarter the wave-length. By using a fork of known number of vibrations, and finding by trial the length of air-column which will respond to the fork, the velocity may be calculated.

EXPERIMENT.

Apparatus.—Resonant tube; tuning-fork; wash-bottle with water; chemical generator; marble; muriatic acid; meter-stick; blocks of wood to support generator; thermometer.

OBJECT.—To determine the velocity of sound, (*a*) in air, (*b*) in carbon dioxide.

MANIPULATION.—The tube being placed upon a firm surface, sound the fork holding it horizontally over its mouth, one prong directly above the other. By means of the wash-bottle, run water into the tube until the point of strongest reinforcement is found. When this point is approached, add the water very cautiously, holding the delivery-tube of the bottle close against the sides of the tube, taking care to wet the sides of the tube as little as possible. Place the meter-stick against the inner wall of the tube, and measure the distance from the level of the water to the top of the tube. Pour out some of the water and repeat, making four or five trials in all. Empty the tube completely, and support the generator on blocks so that the end of the delivery-tube reaches nearly to the bottom of the tube. Pour about 20 cm. of acid into the generator and allow it to run for three or four minutes; then proceed as above. Record the temperature. Tabulate :

Substance used.	Length of air-column.	Average.

CALCULATION.—Average separately the values of the quarter wave-length obtained for air and for carbon dioxide. By substituting these values and the number of vibrations of the fork in the formula

$V =$ No. of vibr. $\times$ 4 (length of air-col. $+ \frac{1}{4}$ diam. of tube)

calculate the velocity of sound in each medium.

APPENDIX A.

GENERAL SUGGESTIONS.

The following suggestions are based upon the author's experience in endeavoring to economize the pupil's laboratory time, and to reduce to a minimum the labor of supervision, correction, etc.

Notes.—The note-book should be the original record of the pupil's work, and should show his failures as well as his successes. He may be allowed to draw his pencil through anything which he does not care to have taken into account by the instructor, but the use of an eraser should be prohibited. He should be compelled to record in the note-book all data as fast as obtained. The use of scraps of paper for notes, and all trusting to memory until after the exercise is completed, should be peremptorily forbidden.

If the notes of all the pupils are uniformly arranged, the labor of the instructor in looking them over is much lightened. A convenient arrangement is the following:

1. *Object.* As in the manual itself.

2. *General Method.* A brief account, in general terms, of the principle upon which the experiment is conducted. It should state (*a*) the body experimented upon, (*b*) the conditions under which it was placed, and (*c*) the proposed observations or measurements.*

3. *Special Method.* A brief description of the manipulation, accompanied by a diagram of the apparatus.

4. *Results.* A careful record of observations of the results of working up numerical data.

5. *Inferences.* Whatever general truths the pupil has gathered from the study of his results.

6. *Remarks.* Under this head would appear statements of errors with means of avoiding them, possible reasons in case of failure, or any other information that the pupil may desire to lay before his instructor.

---

* See Electricity, Exercise 5; Mensuration, Exercise 2.

To get the pupil into the habit of arranging his results methodically, and to make the records more uniform, the arrangement of the data is indicated in many cases in the preceding exercises, and the forms of tables are generally given. A good rule for the arrangement of notes is: Everything written out of the laboratory to be in ink, everything written in the laboratory to be in pencil; all written work to be placed on the right-hand page, all numerical data, calculations, and diagrams on the left-hand page. The handiest form of note-book measures about 8×10 inches, and contains about two hundred pages.

Previous to the exercise the pupil should record, in ink, the object and the general method. From a study of the instructions he should also provide his note-book with blank tables, diagrams, etc., so that the recording of results will require a minimum of time. For example, in Exercise 3, on magnetism, diagrams of the apparatus without the compass-needle may be made, and that may be rapidly sketched in after its deflection has been observed.

It is highly desirable that the instructor get a general idea of each pupil's recorded work before he leaves the laboratory. If practicable, the instructor should look over the laboratory entries as soon as they are made, and if they will pass, the pupil should then draw what inferences he can. These should also be immediately entered and shown to the instructor. Some sort of a check-mark, made at the time to show that this preliminary inspection has been done, will be found convenient when the finished note-books are subsequently corrected. At another time, before or after class discussion of errors, the pupil's individual mistakes may often with advantage be reviewed with him. The books so written up should be handed in at stated intervals and corrected by the instructor.

Essays.—From time to time papers on various completed experiments, beginning with the simpler, may be assigned to the pupils, read before the class, criticised by the members, and afterwards corrected by the instructor. The object of this work would be to train the pupil in explaining briefly, clearly, and minutely the steps by which his knowledge was obtained, to give these steps in their order, and to prepare a clear description of the apparatus used.

Lectures.—In addition to the regular work, occasional suitably illustrated talks on the practical application of physics, particularly in the departments of electricity and heat, are interesting and generally profitable, especially if they can be accompanied by visits to places where the application is made on a commercial basis, as, for example, to an electric-light station or an engine-room.

Preparation of Laboratory Work.—The instructions in the body of the book assume that the parts of the apparatus are set out before the pupils come into the laboratory. The pupils are then expected to connect the different parts, and arrange the whole in accordance with the diagrams. The diagrams are intentionally made in conventional form, with the usual signs for the various pieces of apparatus. Occasionally two different signs for the same piece are used interchangeably. Pupils should be taught to use conventional signs in their note-books.

Apparatus.—Probably the most satisfactory way of getting a laboratory equipment is to buy the standard pieces of apparatus, as thermometers, spring-balances, scales and weights, and to personally supervise the construction of the remainder. For those desirous of following this plan instructions are given in Appendices B and *C*, and, wherever possible, the dealers are indicated from whom such pieces of apparatus, in whole or in part, or their equivalents, may be procured. In some cases temporary substitutes for pieces of apparatus arc described.

Pupil's Preparation.—In order that the laboratory work shall be performed methodically and rapidly, the instructor should satisfy himself that the object, the method, and the manipulation of the exercise have been thoroughly grasped by the pupil before the laboratory work begins. This point is a *sine qua non.*

Teacher's Preparation.—In working through these experiments, for the first time at least, it is highly advisable that the teacher read over the instructions and notes in Appendix D, and perform the experiment himself. In this way he will gain a knowledge of the conditions under which the pupils work, and of their probable mistakes and difficulties, which will make the handling of a laboratory section much easier. The best substitute for this plan would be to get some bright member of the class to work through the experiment and subsequently talk over his experience with him. A trustworthy pupil so prepared is often of considerable aid in acting as an assistant during a laboratory exercise.

Cost.—The author's experience leads him to think that, where the standard pieces of apparatus are purchased, the regular carpentry and metal work done by mechanics, and the remainder of the apparatus constructed with no expense except for raw material, the cost of laboratory equipment would approximate to $15.00 for each pupil in a laboratory section. The details given in Appendices B and C are intended to enable a teacher to make a close estimate for his particular circumstances.

Size of the Sections.—The writer prefers sections of about fifteen, all being given the same exercise with individual pieces of apparatus—except in a few cases where two exercises are worked by half-sections, or by several students together, as indicated in Appendix D. He has handled larger sections; and is of the opinion that if pupils have been thoroughly prepared, fairly successful work can be done with a section of twenty, especially with the aid of a pupil assistant.

Glass-working.—Instructions for glass-working are given in the appendices of most chemical and physical manuals. A method of cutting large bottles and tubes which is not usually given is as follows: Thoroughly soak some common cotton string in water. Wind this wet string smoothly and evenly for a distance of three or four inches on either side of the line that is to be cut, the place where the cut is to be made being left uncovered for a width of about one-eighth inch. In this open space, which should be dry, make a scratch with a sharp three-cornered file, and heat this scratch over the Bunsen burner. Usually, after heating a little while, the glass cracks at the scratch, and breaks away in a clean cut all around. Some practice is needed to be able to tell when the glass has been heated enough.

The Edison Current.—There are a number of experiments illustrating some of the most important modern applications of electricity, which are usually difficult to perform, because they require a comparatively heavy current, and call for an expensive battery outfit, with its attendant evils. For this class of experiments and in other suitable cases the author has used the 110-volt Edison incandescent-lighting circuit. The arrangement of the wiring he adopted is shown in Fig. 103. The positive, negative, and neutral wires are all led into the laboratory.

The positive wire goes first to a rheostat *R*, and then to a double-contact plug-switch *S*. From one half of this switch, *a*, it passes to the lecture-table, to which the negative and neutral wires are also led. From the other half of the switch, *b*, it extends along the wall at the back of the pupils' benches, and the negative runs parallel to it, about eighteen inches below. These form the pupils' circuit. With one contact on the switch the instructor can get the positive with the rheostat, and either the neutral or negative. With the other contact pupils can get the positive through the rheostat and the neutral. At suitable intervals in the pupils' circuit "single-pole cut-outs, *C*, in which lamps are placed, are attached to the positive wire. From these a wire leads to a switch *S′*, and thence terminates in a binding-post *B*. About three inches from this post is placed another, *B′*,

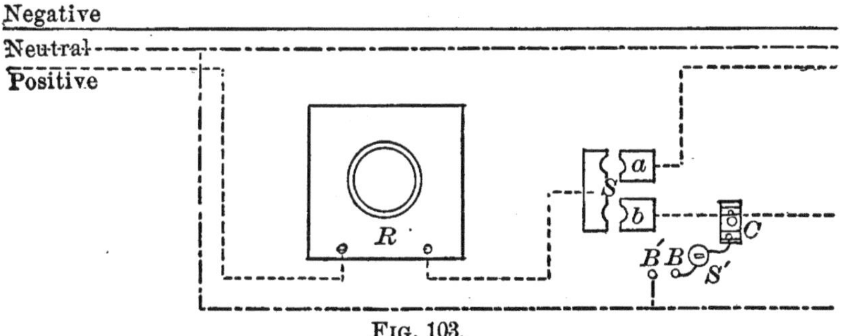

Fig. 103.

connected directly with the neutral wire. On closing the switch the current passes from positive to neutral, through the apparatus. The plus binding-post corresponds to the positive and the neutral to the negative plates of a galvanic cell. The lamps retard the current so that no piece of apparatus will short-circuit and cut out the others, as one of lowest resistance would otherwise do, and so that the turning off and on of the current at one tap will not affect the current at the others. The rheostat need not be resorted to in these cases, the lamps being enough, unless a weaker current than that given by one lamp is desired. The rheostat is then used, but the current through the various circuits will vary if one is turned off or on during the work. Owing to its comparatively high E. M. F. (110 volts) the Edison current cannot be used in all the exercises. The author uses it for Exercises 3, 10, and 11, and for such experiments as are suggested in the appended list.

For the experiments usually performed by the instructor heavier currents can be used, contacts being made between the positive and neutral. The lowest resistance advisable in the rheostat is 7 or 8 ohms, giving a current of 15 ampères down, according to the resistance of the apparatus. The negative wire in the lecture-table may be used when two circuits are needed at once. When apparatus is connected with the negative and neutral wires at the lecture it is placed in series with one or more lamps.

The rheostat may be such a one as is used in Edison lighting-stations, or may be constructed of a number of lamps in parallel. Ten 32 C. P. lamps will give a total resistance of about 11 ohms, and the current may be reduced by using fewer lamps. If the current given by one lamp is still too much, other lamps may be placed in series with it, so as to cut down the current still more.* All circuits should be well protected by safety fuses to guard against short-circuiting. Before using the Edison current for experimental work, the teacher should become familiar with the method of wiring, the various resistances of lamps of varying c. p., etc. In the case of the stronger currents care must be used to regulate the current for the apparatus, in order not to melt or overheat it. It should also be remembered that the heavier currents have a tendency to arc on breaking the circuit; hence all circuit-breaking with heavy currents should be done rapidly. Care should be used to make all connections with sufficiently large wire; No. 16 is small enough.

The above description refers to the Edison "110-*volt*" *incandescent circuit, and to no other.*

Some of the experiments in which the Edison current will be found useful are the following:

Electro-magnets, 16 c. p. lamp;† electric motor (model), 7 to 20 ohms; induced currents, 100 ohms; attraction and repulsion of currents, 7 ohms; thermal effect of current, 7 ohms; chemical effects of current, 400 to 500 ohms; linear coefficient of expansion, 16 c. p. lamp; action and reaction, 150 to 200 ohms; Ampère's law, 16 c. p. lamp; operating model arc-light, 7 ohms; operating model telegraph, 16 c. p. lamp.

Mercury.—The mercury required for amalgamating, that for reversers and mercury-cups, and that intended for other purposes should be kept in separate bottles. Special care should be used to

* See *Scientific American*, January 1891, for the construction of a simple form of lamp rheostat.

† A 16 c. p. light when hot may be taken as having about 220 ohms resistance. Its resistance increases as it becomes cooler.

keep that needed for Boyle's law, indexes, and calibration, clean, dry, and unmixed with other metals.* For amalgamating, it is advisable to have a large tin pan coated with asphalt paint. A table provided with a ledge is useful for work in experiments where mercury is used. When exposed to the air, mercury becomes covered with a scum. To remove this, cut out a circle of letter-paper 10 to 12 cm. in diameter, and fold it twice so as to make a cone of 60°. With a pin punch the point of the cone full of holes, and place it in a dry glass funnel in the mouth of a dry bottle. Pour the mercury into the funnel, keeping its level nearly up to the top of the cone. The scum will keep on top and the clear mercury will run out through the holes into the bottle. Caution pupils about getting mercury on gold jewelry.

TEMPORARY GAS-PIPING.

Where there are not a sufficient number of gas taps for a section, the arrangement as shown in Fig. 104 may be useful. A glass bottle is provided with a cork pierced by several glass tubes. This cork should be well soaked in melted paraffine. One of these glass tubes is connected with the gas tap, the others with the tubes leading to Bunsen burners. In this way one gas tap can be made to serve from two to four students.

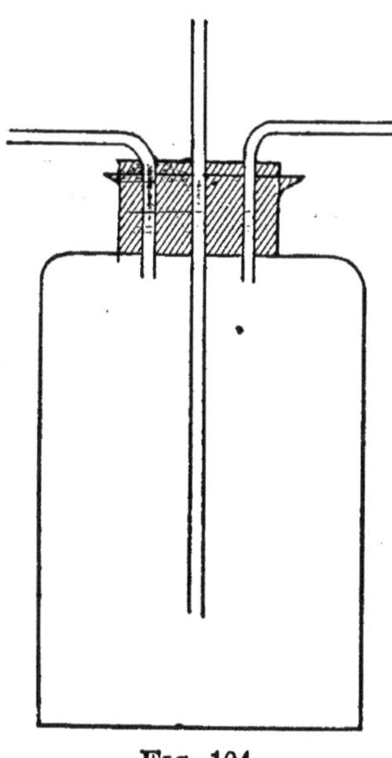

Fig. 104.

If independent regulation is desired, the tube leading to each burner may be provided with a screw clamp † or some similar device. Glass tubing connected with short pieces of rubber tubing, will serve well for temporary piping. Where a branch tap is to be led off, a small bottle provided with a glass tube may be used. By such devices as these, if the laboratory contain but one or two gas taps, it is not a difficult matter to provide enough separate burners.

* For arrangement for keeping mercury clean and dry see Sutton's Volumetric Analysis, p. 413.

† See Eimer & Amend's Catalogue, page 165.

SOME USEFUL BOOKS.

Chute. *Practical Physics.* Boston, D. C. Heath & Co.

Daniell. *Principles of Physics.* New York, Macmillan & Co.

Everett. *Units and Physical Constants.* New York, Macmillan & Co.

Hall. *Elementary Ideas, Definitions and Laws of Dynamics.* Cambridge, Mass., University Bookstore.

Hall & Bergen. *Elementary Physics.* New York, Henry Holt & Co.

Harvard College. *Descriptive List of Elementary Physical Experiments Intended for Use in Preparing Students for Harvard College.* Cambridge, Mass., University Bookstore.

Hopkins. *Experimental Science.* New York, Munn & Co.

Jones. *Physical Problems.* New York, Macmillan & Co.

Kohlrausch. *Physical Measurements.* London, J. & A. Churchill.

Lodge. *Elements of Mechanics.* London, W. & R. Chambers.

Maxwell. *Matter and Motion.* New York, D. Van Nostrand Co.

Mayer. *Sound.* New York, D. Appleton & Co.

Mayer & Barnard. *Light.* New York, D. Appleton & Co.

Pickering. *Physical Manipulation.* Boston, Houghton, Mifflin & Co.

Shaw. *Physics by Experiment.* New York, Effingham Maynard & Co.

Stewart. *An Elementary Treatise on Heat.* Clarendon Press Series. New York, Macmillan & Co.

Stewart & Gee. *Elementary Practical Physics.* New York, Macmillan & Co.

The advertising pages of the "Electrical World" contain illustrations of nearly all kinds of machinery involving the practical application of electricity. Occasional articles in the "Scientific American" also give valuable hints.

APPENDIX B.

LISTS OF APPARATUS.

THE object of this appendix is to place any one undertaking this course in possession of such information regarding the nature and expense of materials required as will most aid him in making an estimate for his individual circumstances and for any particular exercises. The lists are arranged by subject, and each one enumerates, unless otherwise indicated, all the apparatus required by one pupil for all the exercises upon that subject. The notes subjoined give approximate statements of the amount of material required, and suggestions for practicable reduction in total equipment. References have been freely made to certain dealers' catalogues, which it would be well to obtain. The prices given are catalogue prices, and are usually subject to a varying discount, the amount of which can be ascertained of the dealers. Since the same pieces of apparatus are often used in several exercises, items which have appeared in one list are starred in subsequent lists. Constructed apparatus stands in the list as one item, materials being given in the notes. Cross-reference is made by means of the number prefixed to each item. Specifications which might go into the hands of mechanics are given in English units. Wire numbers are B. & S. gauge (see Hall & Bergen, page 372).

ABBREVIATIONS.

E. S. Co., The Educational Supply Co., 9 Hamilton Place, Boston.

E. & A., Eimer & Amend, 205–211 Third Avenue, N. Y.

Gage, A. P. Gage, Boston (Supplementary catalogue, No. 1).

G. & W., Goodnow & Wrightman, 176 Washington Street, Boston.

E. S. G. Co., The E. S. Greeley Co., 57 Dey Street, N. Y.

M. B., Metric Bureau, 146 Franklin Street, Boston.

Ritchie, E. S. Ritchie & Sons, Brookline, Mass. (Catalogue of school apparatus; also special list of apparatus for Harvard College experiments).

W., T. & Co., Whitall, Tatum & Co., 41–43 Broad Street, Boston (1891 catalogue).

MAGNETISM.

For No. 11, *see also Appendix* C.

Apparatus.—1. Bar-magnet. 2. Piece of steel. 3. Carpet-tacks. 4. Pieces of paper, wood, etc.; 3-in. glass rod, E. & A., 8004. 5. Copper tacks. 6. Piece of window-glass. 7. Iron-filings. 8. Iron nails. 9. Block of wood or bottle. 10. Suspension stirrup and thread. 11. Compass. 12. Large darning-needle. 13. Iron and copper wire. 14. Three soft iron nails. 15. Shellac. 16. Glass tube. 17. Wooden clamp.

Notes.—1. Conveniently about $8 \times \frac{3}{8} \times \frac{3}{8}$ in., 35 cts. Substitute: Tool steel, $12 \times \frac{1}{4} \times \frac{1}{4}$ in., G. & W., 10 cts. per ft. Cut bar in two and magnetize with large magnet, or in the field-magnets of a dynamo. Substitute: Large knitting-needle. 2. Knife-blade, knitting-needle. 3. Two sizes. 6. Say 3 × 3 ins. 7. Coarse and fine (E. & A. "fine powder" best for latter; 12 cts. lb.). 10. About No. 18 wire; sewing-thread. 11. Common cheap form, brass, glass top, paper scale about $1\frac{1}{4}$ in. diameter (about 25 cts. each). 13. Say No. 16, 4 or 5 in. 14. Horseshoe nails or clinch nails. 17. Extremely useful. For the form referred to, see E. & A. cat., No. 8222.

ELECTRICITY.

For Nos. 5, 6, 7, 9, 10, 12, 13, 14, 15, 21, 27, 30, *see also Appendix* C.

Apparatus.—1. Five test-tubes: rack or substitute. 2. Bits of sheet zinc, copper (tack), carbon (arc, old battery), amalgamated zinc. 3. Copper and zinc strips. 4. Iron and lead plates. 5. Resistance-box. 6. Reverser. 7. Wheatstone's bridge. 8. Mercury for amalgamation, mercury-cups and reverser. 9. Dilute sulphuric acid and battery fluid. 10. Tumbler cell. 11. Insulated wire. 12. Galvanometer. 13. Mercury-cups and short circuiting wire. 14. Rack with wires. 15. Sensitive galvanometer. 16. Compass.* 17. Iron nail.* 18. Bits of zinc, carbon, wood, glass, etc.* 19. Carriage-bolt. 20. Iron filings.* 21. Mounted coils; 5 yd. g. s. 5 yd. copper, same size, 10 yd. smaller copper (Ritchie E. S. Co.). 22. One large electro-magnet is useful. 23. Binding-posts. 24. Connector. 25. Water. 26. Blocks of wood for supports.* 27. Contact-keys. 28. Bodies for measurement of resistance. 29. Clamp.* 30. Sulphate of copper cell. 31. Bar magnet.* 32. Two electro-magnets.

Notes.—1. E. & A., 8270. W. T. & Co., 2150. Racks, E. & A., p. 219. W. T. & Co., p. 44. Easily made from cigar-boxes. Substitute empty tumbler. 3. Sheet copper, say 2″ × 3″; 20 inches No. 18 insulated wire soldered to upper end, joint protected by asphalt paint, permanent if not handled by wires. Sheet zinc, same size (of hardware dealers), wire as above; attached by punching two or three holes near top, passing one end of wire (insulation removed) two or three times, first through one hole and then through another, and hammering down hard. Plates suitable for this exercise from E. S. Co. 4. Sheet iron and lead, size above; wires attached as above. Substitute large iron nail and strip of sheet lead with wire twisted around upper end. 5. Regular instrument, Ritchie, Gage, $4.00. If calibrated rack with wires, No. 14, is used, but one regular box is needed for the laboratory. See App. C and D. 6. E. S. Co., 60 cts. Easily constructed. 7. Gage, $4.00. Easily constructed. See App. D. 8. For amalgamation, half pound in separate bottle; for cups and reverser, keep another portion dry; rough estimate, half ounce per hole. Chemical dealers. 9. Chemical dealers. Concentrated acid in 8-lb. bottles; 5 cts. per lb., bottle, 25 cts.; 2 ounces. Bichromatic potassium or sodium, chemical dealers; former, 14 cts., latter, 25 cts. per lb. 10. The cell used for Harvard College Experiments would probably do as well. 11. For connections, 15 ft. No. 18 copper wire, electrical dealers, 40 cts. per lb., 150 ft. per lb. 12. Material, wood, 5 sq. ft.; about 40 ft. No. 18 insulated copper wire; three binding-posts. The H. C. form, E. S. Co., will serve. 13. Material, wood, 9 sq. ins; two binding-posts or block 3″ × 3″ × 1″, and wood-screw; 6 inches No. 16 copper wire. 14. Material: meter-stick, see Mensuration, board, 40″ × 6″ (1½ sq. ft.); uprights, 1″ × 2″ × 8″; 7 feet of german-silver wire; 17 feet iron and 50 feet uncovered copper wire, say No. 24 (see App. C.); 6 inches No. 16 uncovered copper wire; binding-post; also one Eng. binding-post if possible; Ritchie E.

S. Co., E. S. G. Co. Iron wire also of hardware dealers; rack without wires, as for Exercise 44, H. C., E. S. Co. 15. Wood for frame; 12 in. glass rod or tube; needles; hair or silk fibre; card; 2 binding-posts; wire, 100 ft., No. 28; No. required, see App. D. 19. Three inch, with nut. 22. Magnetizing model motor, etc.; E. & A. 23. E. & A., E. S. G. Co., 7 cts. up. Total number according to apparatus constructed. 24. Useful, E. S. G. Co., p. 90; Ritchie, E. & A., catalogue No. 5614, 10 cts. Substitute. Large binding-post with hole large enough to carry two wires at once. 27. E. S. G. Co., pp. 310, 326 (four or more old telephone bells). See App. D. 30. Useful in certain cases. See App. D. E. S. Co. 30. Porous cup; E. S. Co. Vessel, sheet zinc, solutions of copper sulphate and sulphuric acid. See App. D. Whole cell, E. S. Co. 32. E. S. G. Co., G. & W.

MENSURATION.

For 5, 9. 13, see Appendix C.

Apparatus.—1. Piece of paper with crosses. 2. Meter-stick. 3. Cardboard circles. 4. Paper for markers. 5. Graduated cylinder. 6. Body of unknown volume. 7. Eight inches fine wire. 8. Four or five rubber bands. 9. Burette or equivalent. 10. Ungraduated jar or equivalent. 11. Glass tube. 12. Bodies for estimation of length. 13. Scales and weights. 14. Avoirdupois weights or 8-oz. balances. 15. Bodies weighing from 1 to 2 oz. 16. Two beakers (1 oz. or so). 17. Saturated solution calcium chloride and $\frac{1}{8}$ sulphuric acid, or equivalents. Any two solutions giving precipitates are suitable. 18. Pencils. 19. Water. 20. Corks. 21. Blocks of wood.* 22. Tumbler.* 23. Test-tube.* 24. Suspension-wire. 25. Wax or paraffine shavings. 26. Bits of stick caustic potash (E. & A.). 27. Burner or equivalent.

Notes.—1. Crosses ruled with sharp pencil, 30 to 50 cm. apart. 2. English on one side, French on the other. Met. Bureau, p. 8. Ritchie, 25 cts. Substitute, form of school rule, 12-inch scale one side, 30 cm. on the other. Met. Bureau, p. 8. 3. Struck off with dividers on stiff cardboard, then cut out with fair degree of care; 10, 6, and 4 in. diam. 4. Strips 1 × 5 in.; writing-paper best. 4. E, & A. 6138 (65 c.). Substitutes may be constructed. For other sizes, see same catalogue. 6. Stone, electric light carbon, vol. 5 or 10 cm. 9. E. & A., 5973. Mohr's burette, price is according to size. 10. E. & A., 6134. 4×1 inch (15 c.), all sizes. Substitute piece of tubing about 1 inch inside diam., 6 or 8 ins. long, lower end closed with paraffine cork and stuck into hole bored in block of wood. 11. 10 in. × $\frac{3}{8}$ in. intern. diam. is good size. E. & A., 6541, No. 7, for example. Provide cork. 12. Books, tables, blocks of wood, etc. 13. Scales, E. & A., 5422* ($1.75). Ritchie, E. S. Co. Weights, E. & A., Ritchie, E. S. Co. 14. As used for weighing letters. Complete equipment, one apiece. Without exercise on friction, one to every three students is sufficient. Obviates need of avoirdupois weights. 16. E. & A., W. T. & Co. 17. In stock bottles labelled "No. 1," "No. 2."

DENSITY AND SPECIFIC GRAVITY.

Apparatus. 1. Bodies for the determination of density. 2. Scales and weights.* 3. Graduated cylinder.* 4. Paper markers. 5. Beaker.* 6. Wire.* 7. Specific-gravity bottle. E. & A., 5683; W. T. & Co., p. 15. 8. Liquids and solids for specific gravity. 9. Apparatus for liquid pressure. 10. Apparatus for specific gravity by balancing. 11. Meter stick.* 12. Funnel. 13. Apparatus for ex. on floating bodies. 14. Apparatus for ex. on atmospheric pressure. 15. Apparatus for specific gravity by barometric columns. 16. 8-oz. balance or avoirdupois weights.* 17. Water. 18. Box. 19. Sinker. 20. Tumbler.* 21. Clamp.* 22. Vaseline. 23. Wire for suspension.

Notes.—1. Rectangular blocks, spheres, etc., irregular body (weighing about 50 grm. each) liquid. 8. Small iron ball, weight 1½ stone, etc. For liquids, alcohol, saturated solutions of calcium chloride, copper sulphate, salt, or equivalents; oil. These may be kept in stock bottles and known by number. 9. See App. D. Materials: meter-stick;* clamp;* wire;* sheet rubber, druggist (4 sq. ins.); thistle tube, E. & A., 6411. W. T. & Co., 2201; rubber tubing (length depends on vessel used), E. & A.; reference-scale; vessel for water, pail or equivalent; piece of board for gauge support; 6 ins. of small glass tube. 10. Material: 2 large glass tubes about 1 meter long, E. & A., 6541, No. 5, or for form given in App. C, 6540, No. 15 and No. 3; support, 3×1 ft.* and baseboard. See App. C.; 8 or 10 ins. fine iron wire or a cork; leather strips. 12. Thistle tube is best. 13. Small test-tube,* corked, containing shot enough to float upright; the whole weighing about 13 grams. 14. Material: Barometer tube, 3 ft., E. & A., 6541, No. 5, one end sealed; 2 lbs. mercury;* feather and wire; meter-stick;* scales* and weights* (up to 500 grams); vessel for mercury bath (small porcelain mortar or equivalent). 15. Material: small glass tubing, E. & A., 6541, No. 4, 6 ft.; bottle; 6 in. rubber tubing to fit glass tubing; small glass plug, size of tubing; fine wire; two vessels for liquids. 18. Starch box or equivalent for supporting scales. 23. Druggists. One bottle lasts a long time.

HEAT.

For 15, 31, 34, 36, 38, *see also Appendix* C.

Apparatus. 1. Metallic rod. 2. Burner.* 3. Clamp.* 4. Wax. 5. Pieces of wood and of glass rod.* 6. Large iron nail.* 7. Two beakers.* 8. Ring-stand. 9. Wire gauze or sand-bath. 10. Potassium permanganate. 11. Thermometer. 12. Ice or snow. 13. Tumbler.* 14. Boiler with connections. 15. Apparatus for testing thermometers. 16. Wooden paddle. 17. Spring balance (or coarse scales and weights). 18. Piece of window glass.* 19. Timepiece. 20. Pint tin pail. 21. Melting-tubes. 22. Stirring-rod. 23. Bodies for melting- and boiling-points. 24. Five test-tubes* and corks, one perforated. 25. Mercury.* 26. Alcohol.* 27. Scales and weights.* 28. Iron ball or equivalent. 29. Cotton string. 30. Calorimeter. 31. Apparatus for latent heat, or substitute. 32. Cardboard screen. 33. Blocks of wood.* 34. Apparatus for expansion of gas, or substitute. 35. Meter-stick.* 36. Apparatus for expansion of liquid. 37. Alcohol. 38. Apparatus for expansion of solid. 39. Two tin cans. 40. Granulated sugar. 41. Powdered sugar. 42. Sand, iodine, copper sulphate. 43. Incandescent lamp. 44. Water. 45. Dividers. 46. White cards. 47. Rubber bands. 48. Three corks. 49. Funnel. 50. Watch or clock. 51. Block of wood.

Notes. 1. Copper, brass, or iron $\frac{1}{2}$ in. $\times$ 1 ft. Wrought iron comes in rods, and may be cut to suitable length, or large electric-light wire may be used. 2. E. & A., 5809, common chemical form with 2 ft. of rubber tubing. Substitutes. Alcohol lamp, E. &. A., p. 165, or glass bottle with perforated cork, carrying glass tube for wick. 4. Common yellow wax or paraffine. 7. Some form of thin glass vessel which can be heated is called for in several of the exercises, and is also used in the form of calorimeter described in 30. The best is a chemical beaker; these can be obtained from dealers in chemical apparatus, and usually come in "nests." E. & A., p. 61, especially 5550 and 5561 *a*, W. T. & Co., p. 15. The next best substitute is a so-called "hot-water tumbler." These can usually be heated on a sand-bath without much danger of breaking. If a metallic calorimeter is preferred, Part 6 of Exercise 1 might be either performed before the class, or with "hot-water tumblers," and the use of the beakers omitted altogether. 8. E. & A., pp. 216, 217; useful form, 8212; substitute wooden clamp,* but not as good. 9. Reduces breakage if flasks are used. E. & A., 8039 and 8442. 10. Chemical dealers. 11. Centigrade thermometer reading to a little above 100° is preferable. Ordinary Fahrenheit thermometers could be used. There is made a cheap form of thermometer which has a paper scale, the whole being enclosed in a glass tube. These are commonly graduated both to Fahrenheit and Centigrade. A very useful form is a Centigrade chemical thermometer reading to 1°, E. S. Ritchie & Sons, $1.00. See also the catalogs of dealers in chemical supplies, and that of the Educational Supply Co. 14. Pint flask, E. & A., 6344 *a*, with safety tube, cork and delivery tube, or pint tin cans with hole in top, fitted to carry cork. Empty "Squibb's ether" cans have a hole so fitted, and could probably be obtained at drug stores. For general form of delivery tube see Fig. 75, p. 134, and Fig. 76, p. 137. (No safety tube shown.) 17. 64-oz. balances best; two or three are sufficient. 24-lb. balances might be used. 21. Time is saved by giving out ready filled. 22. Made of glass rod, E. & A., p. 154. 6 in. glass rod also needed for 5. 23. Paraffine, wax, benzine, ether, alcohol. 25. $\frac{3}{4}$ ounce. 28. Cast-iron with hole drilled through them, Gage, Ritchie. Substitutes, glass stopper, lead bullet, etc. 30. A form which has given good results is made by placing a chemical beaker either inside of a larger one, or in a tumbler, and filling the space between the bottom and sides of the two vessels with excelsior, or some other non-conductor. Another form is a nickel-plated "liquor-shaker." E. S. Co. This is less liable to be broken, but is somewhat more expensive. Bright tin cups or pails would also serve the purpose. 31. Boiler and connections, see 14; ring-stand, see 8; burner, see 2. Material: 5 ft. glass tubing, E. & A., 6540, No. 3; 2 ins. small rubber tubing for connectors, E. &

A., p. 199; 2- to 4-litre glass bottle. The bottles in which acid comes are very suitable. For the bottle might be substituted a tin can fitted to carry a cork in the bottom; 3 ins. large glass tubing, E. & A., No. 15; 3 corks, one for bottle, two for steam trap. At least one litre flask is desirable if students are to fill calorimeter. See App. D. For substitute experiment, ring-stand, boiler, calorimeter, burner, and blocks. 31. 2 × 2 ft. tacked to a baseboard. 34. Material: 4 ft. glass tubing, vessel of 31; vessel for gas (see App. C.); cork; connectors; drop of mercury.* 36. Material, test-tube;* cork; 2 ft. small glass tubing. 37. Specific gravity at average temperature of room previously determined. 38. Material. 1st method: Jacket; rod; wood; 10 cm. scale; 18 in. square brass rod, ¼ in. square; needle; thread; wire for rivets; 2 corks; glass tubing; wire for guys; support for short scale. (For brass rod, see G. & W., p. 76.) 2d method: 2 ft. brass rod, ½ in. diam.; 2 pieces brass plate (heavy), 2 × 2 ins., hardware dealers; clock-face; minute-hand; 6 in. glass tubing; 2 corks; wood for baseboard, etc.; 2 ft. large glass or tin tube, E. & A., 6540; No. 16 is a good size.

DYNAMICS.

For Nos. 1, 5, 8, 10, 11, 14, 17, 18, 21, 22, *see also Appendix. C.*

Apparatus.—1. Apparatus for Exercise 1. 2. 24-lb. spring balance.* 3. Board. 4. Blocks for friction. 5. Dynamometer or 8-oz. balance.* 6. Fish-line. 7. Dividers.* 8. Apparatus for composition of angular forces. 9. Meter-stick.* 10. Apparatus for composition of parallel forces. 11. Apparatus for inclined plane. 12. T-square or plumb-line. 13. Apparatus for pendulum. 14. Apparatus for collision or substitute. 15. Wire of two sizes and two materials. 16. Half spool and screw. 17. Apparatus for elasticity of a solid. 18. Apparatus for elasticity of a gas. 19. Barometer. 20. Mercury. 21. Spiral spring. 22. Hook for Exercise 11. 23. Bodies for determination of specific gravity. 24. Timepiece.* 25. 40-cm. scale. 26. Callipers. 27. Nails. 28. Clamp.

Notes.—1. Material: 10- or 12-lb. weight or loaded box; 8 or 10 ft. No. 18 iron wire; strong screw-eye; cotton string. 2. One, if the last part of Ex. 1 be worked by two pupils. 3. 6 × ½ ft. Two hard-pine matched floor-boards held by cleats on under side are good. May be used also in 11 and 14. 4. 3 × 4 × ⅞ ins. They may be cut from hard-pine floor-boards, planed on one side, screw-eye set into one end of one of them. 6. Smooth cord, strong enough to stand 25 lb. pull. About 25 ft. 8. Three students. Material. Three balances, see 2; one block, see 4; dividers, see 7; cord, see 6; see App. D in connection with the fourth balance. 10. Three students. Material: meter-stick, see 9; cord, see 6; 3 balances, see 2. 11. Two students. Material: balance, see 2; cord, see 6; board, see 8; carriage (skate, Catalogue of J. R. Judd, 1364 Broadway, N. Y.); small box; heavy bodies for loading box; support and rod, or equivalents, see App. *C*; screw-eye; meter-stick, see 9. 13. 3 feet cord fine enough to fit into slot of screw; 2 wood-screws; spool; iron balls of two sizes (see Heat list); for general form of apparatus, see Fig. 91, 14. Two students. Material: board, see 3; meter-stick, see 9; clamps,* 2 ivory balls, Ritchie, E. S. Co.; 50 ft. No. 30 wire for suspension; 8 cards mounted on blocks 2 × 3 ins.; electro-magnets (old telephone bells, E. S. G. Co., G. & W.); current; circuit-breaker, E. S. G. Co., App. *C*, small tacks; board for ceiling, 2 × 1 ft.; screws; putty. If balls are not weighed, scales and weights. Substitute experiment. Three students. Material: Board; 2 meter-sticks; as above; 12 ins. No. 18 spring brass wire, Ritchie; 2 pint tin pails (see heat); cotton string; candle; No. 18 wire for suspension; four screw-eyes; heavy bodies; rough scales. 15. About 25 ft. of each size, breaking-weight under 20 lbs. 17. Material: block, 3 × 3 × 6 ins.; block, 1 × 2 × ½ ins.; block, 3 × 3 × 3 ins.; reading-card with sliding block; elastic bands; meter-stick, see 9; scale pan; 2 ft. No. 28 wire to support same; 2 ins. pencil or equivalent; large and small wood-screws; rubber strip, 36 × ⅛ × ⅛ ins., another, 36 × ⅛ × ½ ins., E. S. Co. 18. Two pupils may work together in order to save mercury. See also App. D. Material: mercury, see 20; meter-stick, see 9; support (see Specific Gravity); 40 ins. glass tubing, E. & A., p. 155, No. 5; 20-cm. scale, see 25; leather strips; reading-cards, see Note 17; small screws; feather and rod. For calibration, scales, weights, small beaker. 19. One in the laboratory. 20. 1 lb. See note 18, and note 14 on Density and Specific Gravity. 21. 4 ft. No. 18 spring brass wire twisted around a broom-stick. 22. See App. *C*. 23. Stones, weights, etc. 25. For 18 useful in 8 also. Made by sawing up a meter-stick. 1 meter-stick required for this to 2 sets apparatus for 18.. Substitute in 18, after calibrating put volume scale on cardboard and attach to short arm. 26. Useful for determining diameter in 13. Substitute, rectangular blocks.

LIGHT.

For 1 and 2, see also Appendix C.

Apparatus.—1. Apparatus for *a*, lens work; *b*, intensity of light; *c*, radiation of light. 2. Rumford photometer. 3. Candles. 4. Light for determination of candle-power. 5. Scissors.

Notes.—1. Material: spectacle lens, 6 ins. focal length; meter-stick;* baseboard;* 3 blocks;* cigar-box;* white paper or cardboard; wire nails; candle; chimney; corks.* Material: *b*, in addition to above, white unruled paper. Material: *c*. in addition to material required for *a*, block with perforated screen; arrangement for screening light. 2. Material: baseboard;* 2 meter-sticks;* 2 blocks;* stiff cardboard for screen; lead-pencil or pen-holder. 4. See Appendix C.

SOUND.

For 1 see also Appendix C.

Apparatus.—1. Sonometer. 2. Middle C fork. 3. Carbon dioxide generator. 4. Muriatic acid. 5. Marble. 6. Resonant jar. 7. Wash-bottle. 8. Supporting-blocks. 9. Nails. 10. Meter-stick.*

Notes.—1. Material: 5 ft. Nos. 28 and 30 spring brass wire,* Ritchie; two 24-lb. balances;* 2 ft. strong cord;* 4 triangular blocks; baseboard,* 4 ft. × 6 ins. 2. Ritchie. 3. Material: wide-mouthed bottle; safety tube:* 2 ft. small glass tubing;* cork.* See Shepard's Chemistry. 4. E. & A., 8-lb. bottles; about 2 ounces per student. 6. Hydrometer jar, 1 ft. × 1 in. internal diameter; E. & A., p. 110, or substitute. 7. Material same as for generator. See Shepard's Chemistry, p. 348.

APPENDIX C.

CONSTRUCTION OF APPARATUS.

This appendix contains instructions for the construction of apparatus. An elementary knowledge of glass-working and the simpler carpenters' tools is assumed. If the work be done by regular mechanics, it is well to remember that they have a tendency to give an unnecessary finish, which increases expense.

General Laboratory Equipment.—It is convenient to keep on hand an assorted supply of cord, iron wire, nails, screws, small boxes, etc. Wire should be kept wound upon spools. Necessary tools are hammer, screwdriver, cutting-pliers, shears, and small hand-drill or brad-awl, and in construction work, a pair of tin-shears, jig-saw, cross-cut saw, soldering-apparatus, small plane, and quarter-inch chisel. If standards are to be constructed, a good balance, with weights up to 1 K., burette, set of litre, half-litre and quarter-litre flasks and resistance-box would be needed. A pair of rough grocer's scales will prove useful. Hard-pine floor-boards make good baseboards, etc.; if tongued and grooved, two or more lengths may be joined together. Shellac, glue, asphalt paint, and a few large bottles for solutions, etc., are needed.

Wire and Binding-posts.—Probably the best plan would be to obtain a pound of No. 18 and of Nos. 28 or 30 insulated wire, and of other wires such quantities as may be required in construction. Binding-posts can sometimes be cheaply obtained from old telephone apparatus. Sufficient contact can usually be got by screwing the binding-post nearly in, taking two or three turns around it with the end of the wire, and then screwing the binding-post firmly on it. It is better, however, to solder the wire to the post.

MAGNETISM.

Compasses.—For magnetic work alone, substitutes for regular compasses are readily constructed. As illustrations, three are given.

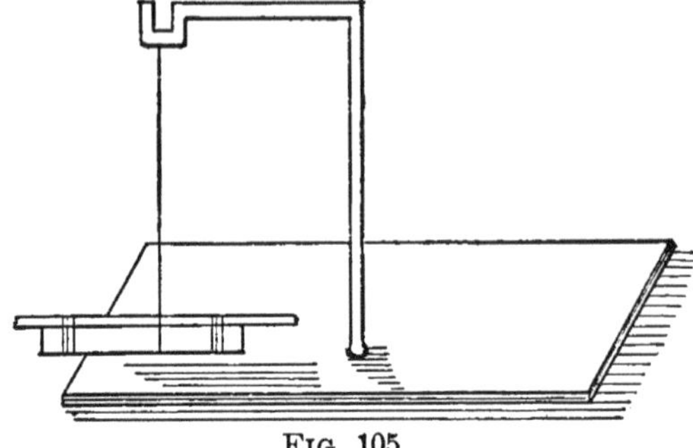

FIG. 105.

I. A support is made of glass tubing or rod bent into the form shown in Fig. 105. This is set into a support of wood about 4 in. square and ¼ in. thick (a piece of a cigar-box for example), to which it may be more firmly attached with sealing-wax. From this support a magnetized knitting-needle 3 in. long is suspended by a hair or fine thread. It is a good plan to weight the needle with a bit of sheet lead tied on with waxed thread.

II. Support of cigar-box wood, as shown in Fig. 106. Base

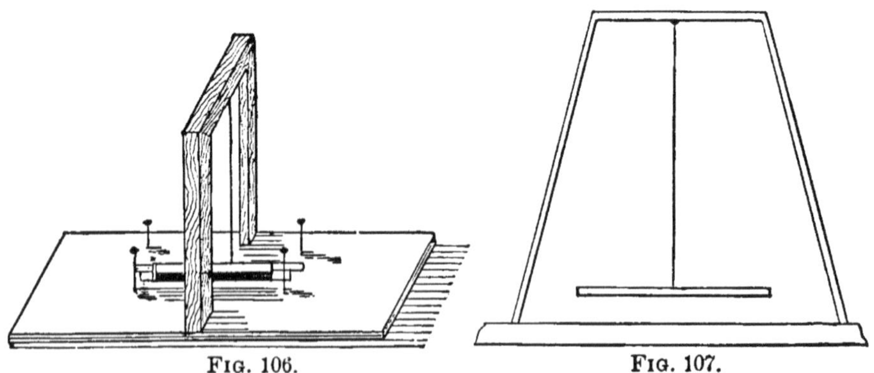

FIG. 106. FIG. 107.

3 × 4 in.; supports 3 in. high. Pins driven into the base prevent the needle from swinging too far. The needle is loaded and suspended as in I.

III. A temporary substitute for a compass may be constructed by suspending the needle from the bottom of an inverted tumbler as in Fig. 107. The suspending fibre is held to the glass by shoemaker's wax.

ELECTRICITY.

Resistance-boxes.—If one regular box is available, others can be constructed by mounting in a cigar-box insulated German-silver wire of suitable lengths and sizes. The capacity of this substitute may be made less than that of the regular box, but bodies of resistance within its capacity must be assigned for measurement. The resistance of the wire per cm. is determined by measuring the resistance of four or five lengths and averaging these values, and lengths required to form coils of any desired resistance are then measured off. Contact can be secured by switches or by some other device, such as the "post-office form." * The points for the switches can be made of round-headed McGill's paper-fasteners, set into holes bored in the wood, the ends bent apart below, and the coils soldered on as in Fig. 108.

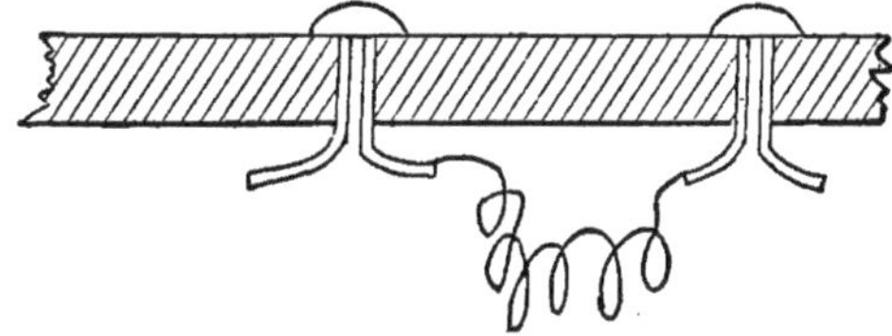

FIG. 108.

The rack with wires described on p. 231 may be converted into a resistance-box by determining the resistance of 1 cm. of the G. S. wire. Get the resistance of a number of lengths, and take their average value for the resistance of 1 cm. Enter this value upon a card attached to the instrument. The total resistance is found by multiplying this value by the number of cm. in circuit. A scale reading directly in ohms could also be easily made and placed upon the apparatus. The capacity may be increased by using more G. S. wire.

Current-reverser.—This instrument is constructed on the same principle as the mercury-cups.† In a block of hard pine or other suitable wood, $3 \times 3 \times \frac{7}{8}$ inches, bore four half-inch holes one half inch deep, whose centres are at the four corners of a square. Cut two pieces of No. 16 insulated wire two inches longer than the diagonal distance between the centres of the holes, and remove the insulation from each end for about an inch. Twist the

* See page 133, Stewart and Gee, vol. 2. † See page 230.

centre of one piece around the centre of the other for a short distance and bend the ends into the general form of an X, as in-

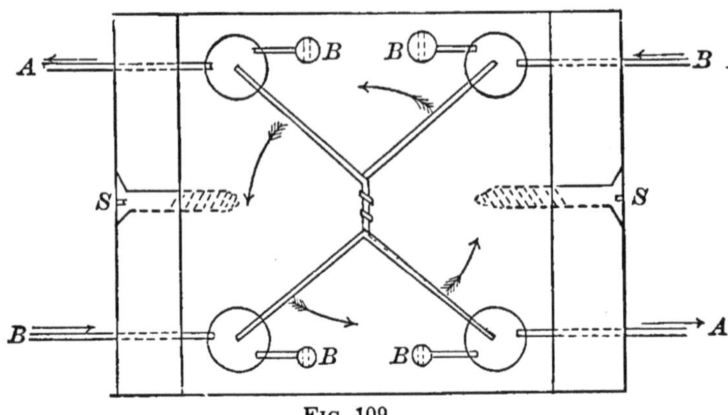

FIG. 109.

dicated in Fig. 109. Turn the uncovered ends down at right angles so as to rest in the holes. Either binding-posts or the

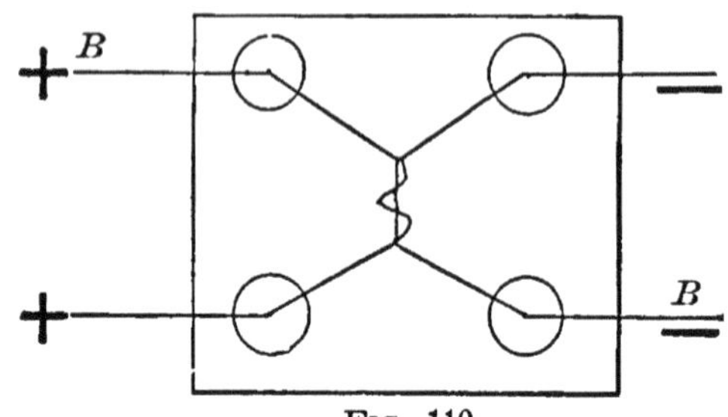

FIG. 110.

substitute may be used ; the latter is just as good, as the instrument is always inserted in the circuit between others. When

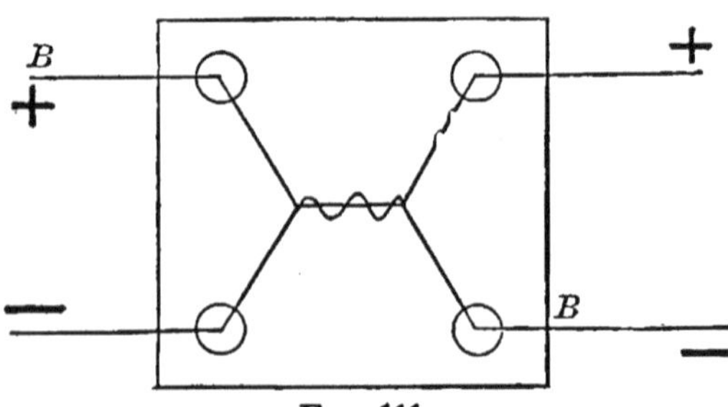

FIG. 111.

the wires leading to the instrument are connected with two diagonal holes, and the wires leading from it with the other two, on filling the holes with mercury, and inserting the "X," the current will pass as shown in Fig. 110: on lifting the "X," turning it half round, and replacing, it the current will be reversed, and pass as shown in Fig. 111. The ends of the wires should be scraped bright before using, or else amalgamated. When placing the "X" in the mercury, it should be moved around a little before the contact is considered good. The current is reversed for each rotation of 90°.

Wheatstone's Bridge. (See Fig. 31.)—Make the base-board about 14 in. square. Use No. 16 wire or strips of heavy sheet brass. No. 18 wires soldered on at the points *b* and *c* for galvanometer, and at *e* and *f* for resistance-box connection, would reduce the number of binding-posts. The posts at *a*, *d*, *g*, and *h* could also be replaced by the substitutes, as in the construction of mercury-cups.

Battery Fluid.—One formula is: Water, 900 grams; pulverized potassium bichromate, 120 grams, or sodium bichromate, 104 grams; concentrated sulphuric acid, 200 grams, or 100 cu. cm. Put half the water into a vessel of as thin glass as possible, add the potassium bichromate, then the acid, with constant stirring until the liquid becomes quite warm. If it gets too hot, reduce the temperature by adding more water. Keeping the liquid hot and stirring it facilitates the solution of the solid. The ease with which the temperature can be reduced with water decreases the liability of breaking the vessel. The solution may be kept in a stock bottle and used until quite green. At the close of each laboratory exercise the fluid in the cells should be poured back into the stock bottle, and the cells refilled for the next exercise or section of the class.

Tumbler Cell. (See Fig. 13.)—This is a bichromate cell. The carbon plates may be made by sawing up the carbons used in the Westinghouse alternating-current arc-light. These come in flat strips, 5 or 6 inches long, about one third of an inch thick, and 2½ inches wide. Carbon, plates may also be obtained of dealers. To cut the carbon, fasten a saw having quite fine teeth upside down in a vise. Holding the carbon in both hands, rub it back and forth rapidly on the saw, bearing down enough to make the saw cut. When partially sawed through, the carbon is liable to break suddenly at the cut, and care must be taken to avoid lacerating the hands on the saw. Drill a hole through the carbon near the top. Pass a piece of No. 18 wire, the insulation of which has been removed for 2 or 3 inches, through the hole, and twist the end firmly about the main wire until it is pulled up hard against the carbon. Dip the top of the carbon in melted paraffine to a depth of about half an inch. It is convenient to wind the connecting wires into a spiral. The zinc plates may be constructed as follows: To make the mould, cut out of a cigar-box cover a piece the size of a carbon plate; lay a square of sheet-iron about 4 × 4 in. upon a piece of hard-wood board, place over this the cigar-box cover with the cut-out portion removed, and screw or nail the three firmly together. Scrap-zinc or old battery-zincs may be melted over a Bunsen burner, or better, over a gas stove. After the zinc is melted, it should not be left in the ladle very long, as it oxidizes rapidly. Take a piece of battery-wire, say No. 18, from 12 to 18 inches in length, remove the insulation for about 2 inches at one end, and bend that end in the form of the letter S. Place this end in the mould, as indicated in Fig. 112, so that it

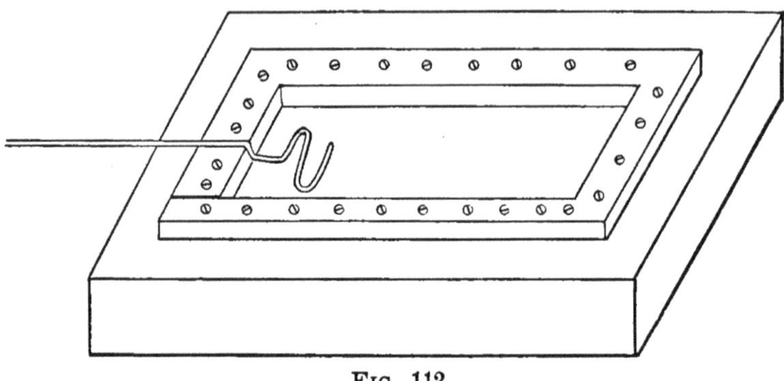

FIG. 112.

does not touch the bottom of the mould, and so that it will be covered by the zinc. Pour in the melted zinc in a thin stream, being careful that no scum passes, until no more can be added without overflowing. Leave it in the mould until nearly cold. Tip the mould upside down, give it a few raps on the upper side, and the zinc will fall out. In handling the plates do not lift them by the wire, as it is liable to break off. Size of plates depends on size of tumbler, but plates of App. B, 3 and 4, must be same size.

Soak in melted paraffine some pieces of cigar-box about one quarter of an inch wide and as long as the plates are wide. Lay the carbon plate on the table; put one of these pieces of wood near one end and another near the other end; place the zinc plate, previously amalgamated, on top, and slip a strong rubber band over the two plates to hold them together. Insert the plates in the tumbler. In adding the liquid, take care that it does not get as high as the points where the wires are attached to the plates. Keep the plates out of the solution except when in actual use. Directions for the construction of a similar form of cell, in which the ordinary round arc-light carbons are used, will be found in the *Scientific American* for May 10, 1890.

Galvanometer. (See Fig. 16.)—This galvanometer is constructed as follows: The hoop *H* is a circle of wood 14 in. in diameter and 1¼ in. thick. Out of it is cut a semicircle as shown, 11½ in. in diameter. The outer edge is grooved to hold the wire. The support *D* is a piece of wood 10 × 6 in. and 1¼ in. thick, having cut in it a slot to receive the hoop. The base-board *S* is 12 × 10 in. and 1¼ in. thick. Starting at lowest point of the hoop and leaving an end of three or four inches, wind the wire (No. 18, "Office Wire") once around, leave a loop, and wind it around nine times more. If the hoop is grooved the coil need not be secured. If the hoop is not grooved, the coil may be bound to the rim by pieces of string passing through holes bored in the hoop near the rim. Place the hoop in the support, having pulled the ends of the wires up through the slot, in which a small groove may be cut if needed. Fasten the hoop in place by wedges or glue. Secure the support to the base-board by brass or copper nails or screws. Attach the wires to the binding-posts, and shellac the whole to prevent warping. The above dimensions give an instrument heavy enough to stand firm, and not easy to break. A cheaper form could be constructed by making the hoop from the top or bottom of a salt-box, securing the coil in place by strings, and making the other parts in proportion. The base-board should be thick and heavy. As the hoop would be thinner, a circle of card-board as large as a silver dollar, fastened in the centre of the semicircle cut in the hoop, would form a good support for the compass. The only expense of this form would be a few cents for the wire, and the cost of the binding-posts. A good base for this form is a cigar-box, such as contains 50 cigars, filled with small stones and nailed up. A substitute for a compass can be made by suspending a magnetized needle from the hoop, allowing it to hang in the centre. It is not as satisfactory as a compass.

Mercury Cups. (See Fig. 113.)—Constructed out of blocks of hard-wood (hard pine is good), 3 × 2 in. and $\frac{7}{8}$ or 1 in. thick. Near one end bore two holes, *H*, half an inch in diameter and half an inch deep, and near the other ends place two binding-posts, *B*. Shellac the block and the insides of the holes. Connect the binding-posts with the holes by two pieces of No. 16 copper wire, *W*, one end of each piece being fastened to a binding-post, and the

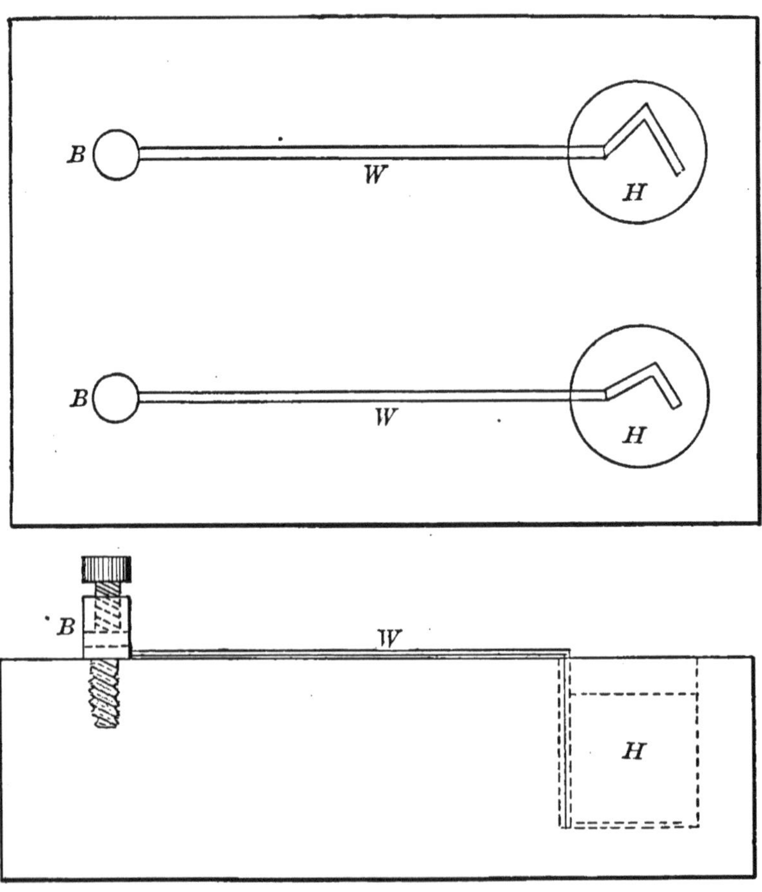

Fig. 113.

other running down to the bottom of the corresponding hole, where it is bent into the form of a flat spiral. If the wire is insulated, the insulation must be removed from the part in the hole, and the wire scraped clean and bright. For use, fill the holes two-thirds full with mercury.

The only expense connected with this instrument is that of the binding-posts. These may be dispensed with by the arrangement shown in Fig. 114. Attach a strip of wood $\frac{1}{2} \times \frac{1}{2} \times 2$ in. to the top of the block by a single screw. Pass under this strip the wires by which the instrument is placed in the circuit by loosening the screw a little and then turning it down firmly, scrape the ends bright, bend as directed above, and insert them in the holes.

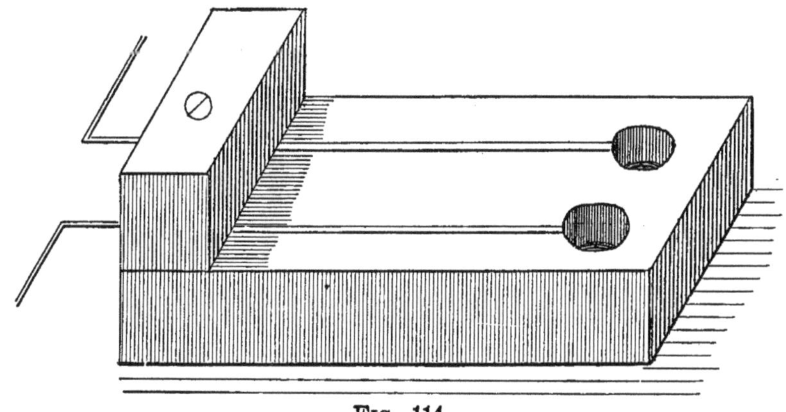

FIG. 114.

Rack with Wires. (See Fig. 28).—The base-board *A* is 5 or 6 in. wide, and of such length that the centres of the uprights are a meter apart. The ends of the meter-stick rest in notches cut in the uprights. Small tacks are driven into the uprights, at intervals along vertical lines dropped from the ends of the meter-stick. From the binding-post a naked German-silver wire is carried to the lowest tack on the other upright and back to the first upright, as shown. A piece of No. 18 uncovered copper wire runs up from binding-post *a*, and to it are attached the ends of the naked iron and copper wire, as indicated. The number of lengths of these depends upon the size of wires. For the sliding contact, the end of a piece of No. 18 copper wire may be hammered out flat, and then bent into the form of a hook. An English binding-post is best.

Sensitive Galvanometer. (See Fig. 115.)—The frame is of wood about 4×4 in.; the support for the needle may be made of wood or glass tubing; and the needle itself a piece of magnetized knitting-needle short enough to swing freely in the frame. A piece of writing-paper is cut in the form shown, and the needle stuck through the lower part. A second needle, for a pointer, is stuck through the upper part. The two needles should be about one inch apart and the lower one only should be magnetized. The frame is wound with wire as shown by the dotted lines, with as many turns as desired. Space is left in the middle to admit the

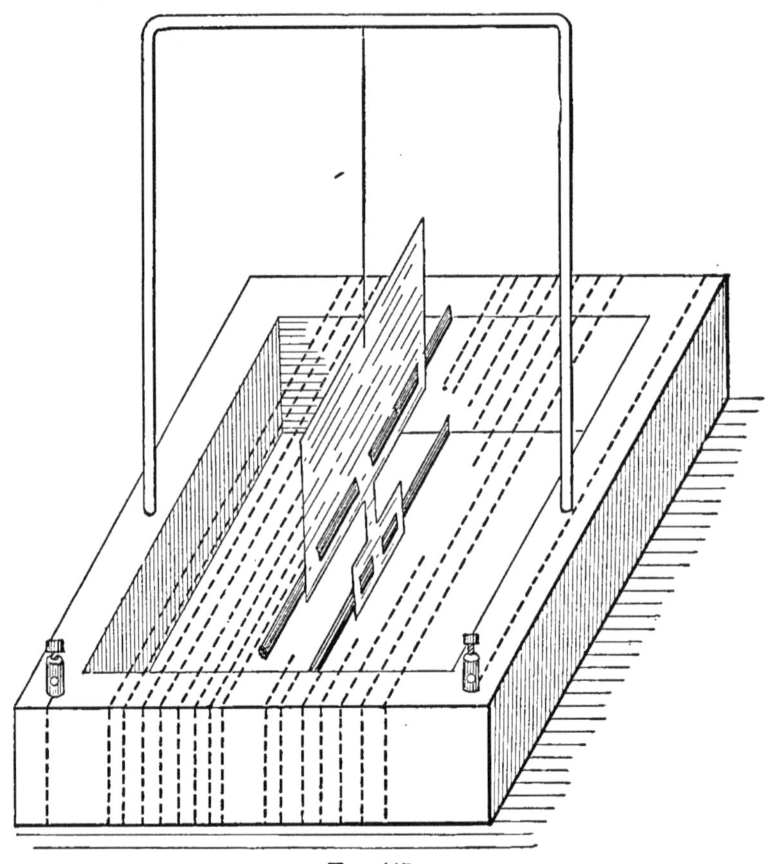

FIG. 115.

needle, which is then suspended by a hair, or a thread of untwisted silk, so that it swings freely within the coil. If desired, a scale may be marked on a circular piece of card-board, and laid on the coil, but it is not necessary. In the diagram the card shown above the indicator swings with it and serves to show the motion to a class. If this were used, the indicating-needle would not be needed. The whole may be covered with a bell-jar, or a large glass bottle whose bottom has been cut off.

Coils of Wire.—These are made of cotton-insulated wire measured off in the lengths indicated in Exercise 4, wound into a coil on two fingers and attached to a piece of stiff card-board by paper-fasteners or wire passing through holes in the card and around the coil. The number and length of the wire are marked on the card, which may then be shellaced to preserve it. The ends of the coils are left free for two or three in., and the insulation removed from these ends. For the 5 and 10 yd. coil, the 10 yd. are looped in the middle, and then wound so as to leave the loop projecting for a few in. The insulation is removed from the loop, and it is twisted together ; or, if preferred, they may be wound upon spools instead of being mounted on cards, the loose ends being drawn through holes bored in the spools. The numbers given in Exercise 4 need not be strictly adhered to, provided the cross-section of the german-silver wire corresponds to that of the same length of copper wire, that the 5, 10, and 20 yd. of copper wire are of so small a cross-section as to make the changes in the galvanometer readings due to changes in lengths sufficiently marked, and that there is considerable difference in the cross-section of the two coils of copper wire 5 yd. long.

Dynamo and Motor. (See Figs. 116, 117.)—The field-magnets are provided by a large electro-magnet.* The armature is a gramme ring, constructed as follows : A circular disk of wood, whose diameter is about one inch less than the distance between the poles of the electro-magnet used, is tacked to the table, and enough soft No. 16 iron wire wound around its rim to form a coil about three eighths of an inch in diameter. The coil, held together by pieces of string tied around it at intervals, is slipped off the disk and wound spirally with a narrow strip of brown paper. It is then wound evenly with No. 18 "office wire" in four equidistant coils, whose ends do not meet by about one inch.† The ends of wire in each coil are left from 4 to 5 inches long. One or more layers may be wound on each coil, as desired. The paper covering between the coils is removed, so as to show the construction of the ring. A wheel or circular disk of wood is mounted as shown in Fig. 117. The axle is made of a wire nail,

*For full instructions for making a simple motor, or an electro-magnet, without special tools, see Hopkins's "Experimental Science."

†This is most conveniently performed by winding the office wire on to a spool, and then using this spool as a shuttle. If more than one layer has been used in the coils, the wheel may need cutting down where the coils come.

with its ends filed sharp. The upper end bears on a brass screw, whose lower end has been filed flat and slightly bored with a drill. The screw is inserted in the cross-piece attached to the base-board. The lower end of the nail rests in a slight depression drilled in the head of a copper rivet. This forms a bearing with little friction. The wheel may be removed by raising the screw. The ring is crowded on to the wheel and secured by pieces of string passing around it and through holes near the rim. A

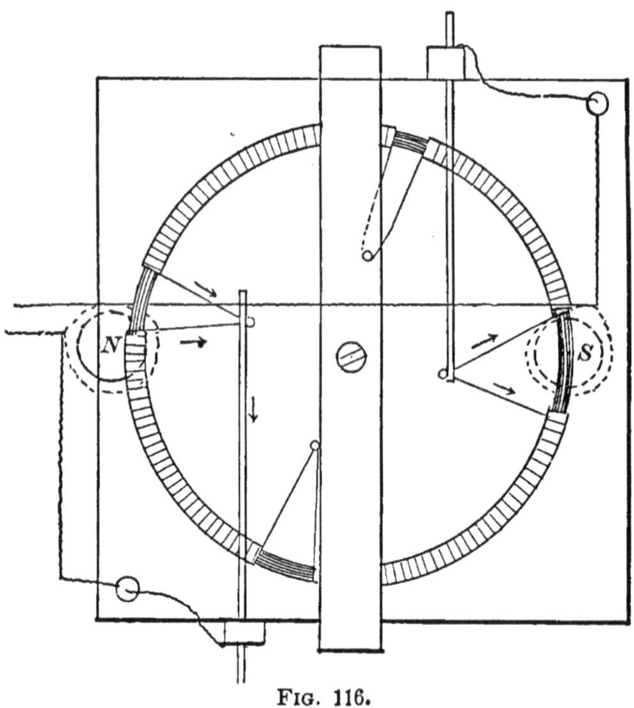

FIG. 116.

circle whose diameter is about half that of the wheel, and whose centre is the same, is struck off on the wood; and four holes, a little smaller than four wire nails, are bored through the wood at points on the circle 90° apart, and corresponding to the ends of the coils. The ends of the two adjacent coils, freed from insulation from the point, where they enter, are drawn down into the same hole, tight, and a wire nail driven in as a wedge. The two wires and the nail, which form a sufficient contact, are cut off even on the under side, and the nail is allowed to project about half an inch on the upper side. Proceed in the same way with the other coils. The "brushes" are two pieces of No. 16 copper wire held by wooden uprights, and connected there by means of No. 18 wire, with the two binding-posts. These brushes should be adjusted so as to press gently against the nails, without rubbing on the wood when the wheel turns.

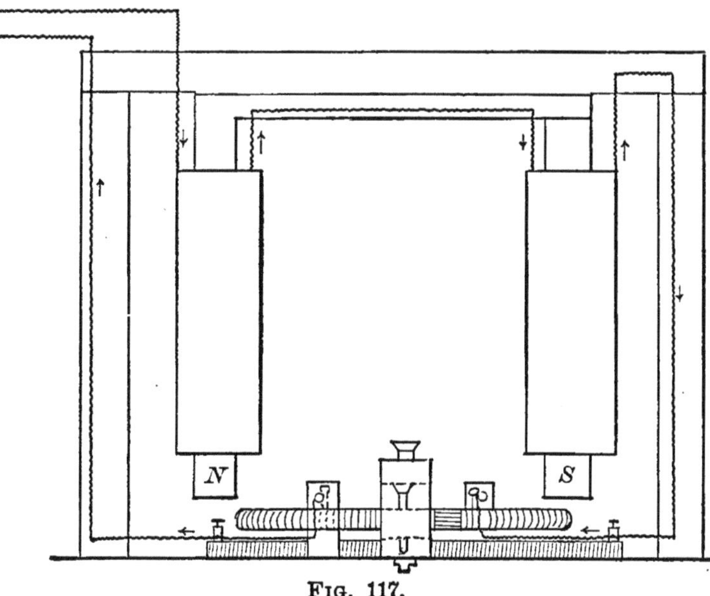

FIG. 117.

This armature is then mounted under the magnet, as near to it as possible without touching, and connections made as indicated.

This contrivance has all the essential parts of a dynamo or motor. On sending through it sufficient current the wheel will revolve, the motion being varied by varying the strength of the current, and reversed by reversing the current through the magnet or the ring. If used as a dynamo, the current is sent through the magnet only, the brushes are connected with a galvanometer, and the wheel is rotated by hand. This contrivance is intended merely to illustrate the construction of the dynamo and motor, and does not develop power. It shows the construction clearly, which a regular motor does not.*

* In the form used in the author's laboratory the wheel was made of the cover of a round salt-box, the ring of 7 cents' worth of iron wire, and the magnet was that referred to on page 216, note 22. The Edison current was used.

Contact Keys.—These may be constructed of sheet-brass about $4 \times \frac{1}{2}$ inches, one end turned up for a handle. A hole is bored through the other end. Through it is passed a machine-screw, one end of a piece of No. 16 wire is twisted around the screw between the brass and the wood, and the screw set firmly in, best by a nut on the under side of the base-board. For an open circuit key the brass is bent so that its free end is about $\frac{1}{4}$ inch from the base-board. Directly below is placed a brass machine-screw with the top filed flat, to which a wire is attached. Both wires may be led to binding-posts, or a length of 30 or 40 cm. may be left permanently attached. This key stands at an open circuit. On depressing the key the contact is completed. The circuit may be maintained either open or closed by leaving the screw so loose that the strip may be rotated, bending the strip slightly down, and pushing the end of the strip off and on on the machine-screw.

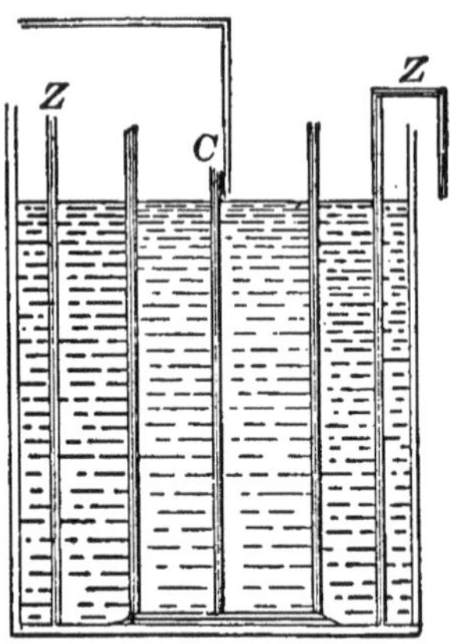

FIG. 118.

Sulphate-of-copper Cell. (See Fig. 118.)—The containing vessel is made by cutting off the upper portion of a flat-bottomed bottle whose diameter is an inch or so greater than that of the porous cup. The zinc-plate *Z* is cut out of sheet-zinc, a projecting strip being left, and rolled around the porous cup into the form of a cylinder, the projecting strip being bent over and having the leading-wire attached to it. For the copper plate *C* one or two of the copper strips used in Exercise I can be used. The porous cup is filled with a saturated solution of copper sulphate, and the zinc plate is immersed in the solution of sulphate of zinc.

MENSURATION.

Measuring Vessels.—Substitutes for burettes, measuring cylinders, etc., may be constructed as follows: A tube is corked at one end and some mercury poured into it. Air-bubbles are carefully removed by means of a feather and wire, as in the experiment on Boyle's Law. The tube is set upright and the height of the mercury-column carefully measured. The mercury is weighed, and its weight divided by 13.6 equals the volume. Several determinations are made, using various quantities of mercury, and the volume in each case when divided by the height of the mercury-column gives the cross-section in sq. cm., and

$$\frac{1}{\text{C. S.}} = \text{length on the tube corresponding to 1 cu. cm.}$$

The simplest way is to attach the tube firmly to a meter-stick, and prepare a table, giving the readings on the meter-stick corresponding to the various volumes desired. Or, a scale-reading directly in cu. cm. may be ruled on paper, glued on the back of the tube, face to the glass. When dry, the back may be well shellacked to prevent injury from water. The mercury used should be clean and dry. The graduation may be performed with water, but mercury is much more desirable, because any errors in weighing only produce one-thirteenth the error in the volume. Graduates accurate enough for the purposes of this book can be obtained by weighing the mercury to a centigram. The chief errors arise from the uneven bore of the tube, and from failure to measure accurately the height of the mercury-column. This method may be used in calibrating the short arm of the tube used for Boyle's Law, and also in graduating flat-bottomed chemical flasks.

In Exercise 3 a substitute for the burette may be constructed as follows, provided one good burette can be obtained: A piece of glass tubing of 8 or 10 mm. internal diameter and about 60 cm. long is provided at its lower end with a cork through which passes a glass tube. This connects at its lower end with a piece of rubber tubing provided with a glass ball, as described on page 73. The whole is firmly lashed to the edge of a meter-stick. A little water is poured into the tube, the reading of its level noted on the scale, a measured volume of water run in from a burette, and the reading of its level again noted. The difference in the scale-readings gives the height occupied in the tube by the known volume of water, and the distance on the scale corresponding to 1 cu. cm. can be readily calculated. Several determinations should be made, the average volume taken and marked on a card attached to the apparatus. When used, the scale-readings are taken before and after running water out of the burette, and the difference in the readings divided by the distance on the scale corresponding to 1 cu. cm. gives the volume. The preparation of such an instrument might be well given as a modification of Exercise 4. Increased accuracy in reading can be obtained by some such device as that described on page 185.

Scales.—A cheap and useful form are brass hand-scales with a 6-inch beam. They are sensitive to about .01 grm., and may be mounted on a five-pound starch-box, as shown in Fig. 119, the frame being of such height that the pans hang about half an inch above the upper surface of the cover. This frame is constructed of hard pine 2×2 in. The scales may be suspended from the cross-piece with a screw-eye, or, if there is a tendency to turn while weighing, a block of wood, as *A* in Fig. 120, provided with a slot to receive the top of the scale-support *B*, which is held by the pin *P*, may be substituted. For suspension weighing, the cover is pulled out and laid crosswise as a support for the weight-pan, as indicated by the dotted lines in Fig. 119.

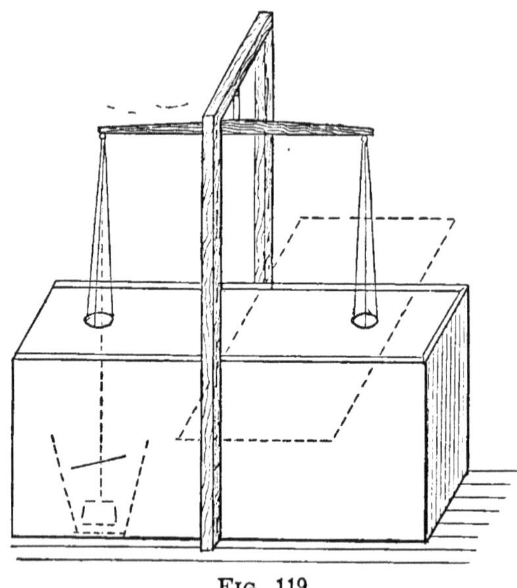

FIG. 119.

A body too large for the pan may be conveniently weighed by suspension. For specific-gravity work a tumbler of liquid may be placed as indicated by the dotted lines in the figure. For better observation in this case part of the side of the box may be removed.

Another good form, made to order for the author's laboratory,* is shown in Fig. 62. The beam is supported upon a sharp edge by a brass post. The lower end of the post is set into the side of a starch-box. Either form is fitted for suspension by piercing a hole through the centre of each pan. In the second form a half-inch hole is also bored through the wood directly below the left-hand pan. This form is a little more expensive and more convenient than the first. The capacity of both forms is from 200 to 250 grm., and their sensitiveness is nearly alike.

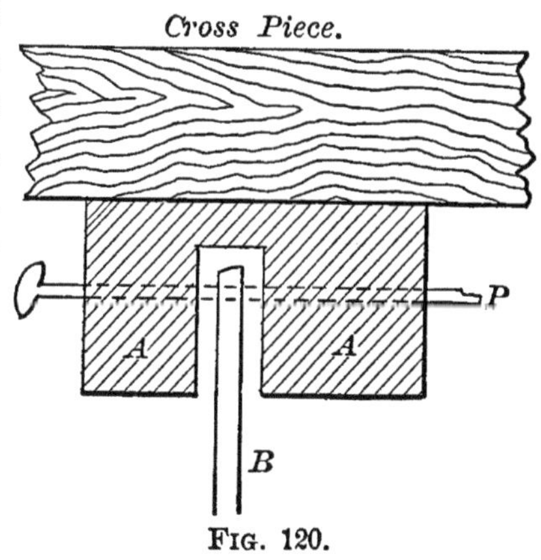

FIG. 120.

Substitute Balances.—Balances may be constructed with a wooden beam of tough wood and a wooden post attached to a suitable base-board. The pans, of tin (spice-box covers will answer) or stiff card-board, may be suspended by fine wire or trout-line. The beam should have some form of knife-edge bearing. One device is indicated in Fig. 121. The post *P* is provided

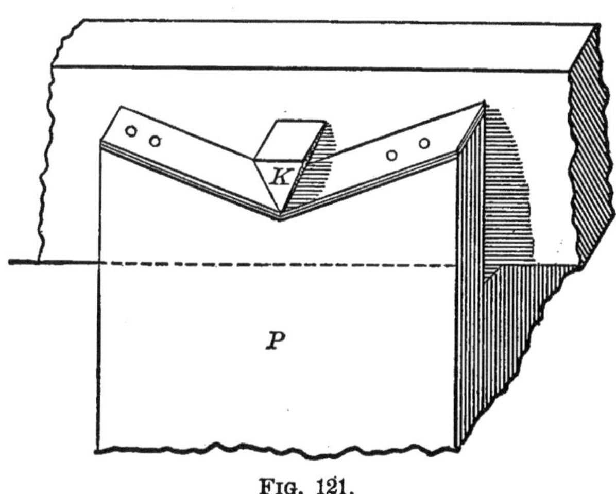

FIG. 121.

with a slot in its upper end about $\frac{1}{4}$ inch wider than the beam, and deep enough to permit the beam to move freely in a vertical

* By Ritchie.

plane. The projecting ends on each side of the slot are cut in the form of a flat "V," with the angle slightly rounded. Pieces of sheet-brass are fitted to these notches and fastened in place. In the beam is inserted transversely a knife-edge, *K*, made of a piece of sheet-brass, whose lower edge has been filed sharp where it will bear on the sheet-support. Care must be used in weighing that

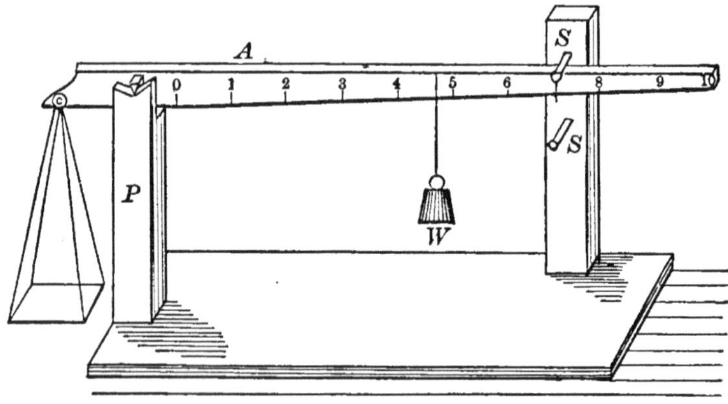

FIG. 122.

the beam does not rub against the sides of the slot. Increased delicacy could be attained by projecting beyond the support one end of the knife-edge and attaching to it a long pointer reading on a scale upon the lower part of the post. In using these scales it would probably be best to weigh by substitution, except when ratios only are required.

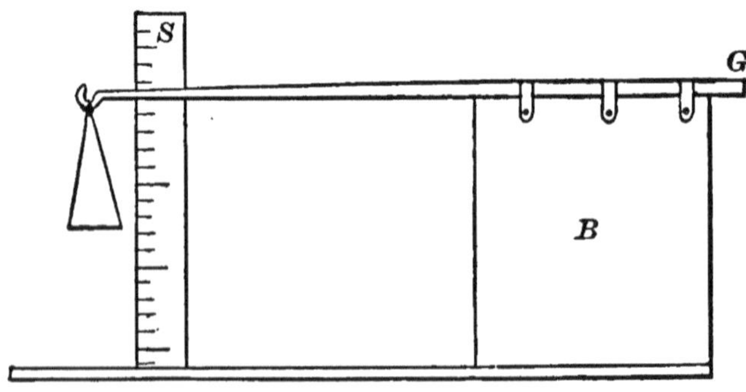

FIG. 123.

A form of "Post-office" balances is shown in Fig. 122. The beam *A* is mounted on the post *P*, and carries a sliding-weight *W*. The play of the beam is limited by the stops *SS*. The scale on the beam is spaced off by finding, by trial with various weights, the average space corresponding to the unit of weight used. These spaces with subdivisions are then laid off on the beam. This device, mounted on a box, would be very good for specific-gravity work. The capacity could be varied by using several weights which were multiples, etc. A contrivance for determining small weights is shown in Fig. 123. A block of wood, *B*, is attached to a base-board. A piece of glass rod, *G*, is drawn out at one end and fastened to *B*. The end carries a light scale-pan, and the deflections caused by various weights may be read on the scale *S*. The deflections caused by known weights are found by trial, and the scale spaced off accordingly. This instrument may be made very sensitive, though its capacity is limited, by drawing the rod out quite long and fine. Within its limits it is quite accurate.

Weights.—The weights referred to in this book are sets of 20 grm. to 0.01 centigram in a box. These are supplemented by weights of 50, 100, and 200 grm., made of lead by aid of a sensitive balance and a large set of weights. For other purposes, elasticity, etc., weights of 500 and 1000 grm. were also constructed of lead, by the method given below. Provided one set of weights and a fairly sensitive balance* are available, sets of weights may be readily made, those of 5 grm. and over of sheet-lead, 1 to 5 grm. of sheet-brass or even sheet-iron, and the decigram weights of thin sheet-brass or wire. For the centigram weights wire of iron or copper, or better, sheet-aluminum, may be used.† The most convenient method for sheet-metal consists in weighing several rectangular pieces of known area, and thus determining approximately the average weight per sq. cm. The area to give the required weight is computed, and a piece a little in excess of this is cut off, transferred to the balance, and carefully worked down by scissors and file to the correct weight. If it is to receive a distinguishing mark, apply this before obtaining the exact weight. At the last, great care must be used not to cut off too much at once. The method for wire is the same, the length per grm. being obtained, and calculated lengths for the

---

*Home-made will do. The "glass-rod" form will do well for the smaller decimals.

† This can be obtained in sheet or wire of dealers in chemical supplies, and, owing to its lightness, weights of small denomination are of some size.

desired weight being measured from that. When one approximate weight is obtained, it may be used as a pattern for others of the same denomination. For more on this subject see Pickering's "Physical Manipulations, page 49."

DENSITY AND SPECIFIC GRAVITY.

Specific-gravity Bottles.—A light salt-mouthed bottle is used, holding about 100 cu. cm., and provided with a ground-glass stopper. Such a bottle weighs about 100 grams, and is marked, by means of diamond ink, with the words "Specific Gravity." A number is placed upon the bottle and also upon the stopper. This prevents the use of the bottles for other purposes and the interchanging of the stoppers. Somewhat less accurate results could be obtained by the use of small medicine-bottles, the bottle, being filled in each case to the level of a ring scratched around its neck with a file.

Apparatus for Liquid Pressure. (See Fig. 63.)—The support for the gauge bulb is a block of wood, E, 4 × 5⅞ in., out of which is cut a rectangular opening ¼ in. wider than the bulb, and 3½ in. deep. Across the bottom of the opening so formed is stretched a wire held by two tacks. To this wire the gauge is attached at two opposite points by pieces of wire, which are passed around the flange, twisted together, then around the wire, and the ends twisted and cut off. The meter-stick is attached to the block, with its lower end even with the wire. The rubber tube should be sufficiently long to permit the free movement of the gauge. The tube is connected at its outer end to a piece of small glass tubing 6 in. long, which is tied to a meter-stick and which carries a globule of mercury as an index. Several pieces of rubber tubing may be connected by short pieces of glass tubing, but all joints should be well wired. The vessel must be large enough to admit of the gauge being immersed to the depth of 6 or 7 inches. An ordinary water-pail is just the thing. The gauge is constructed as follows (Fig. 124): The tube of a "thistle-tube" is cut off about ½ in. below the bulb. Over the mouth of this is drawn, not very tightly, a piece of thin sheet-rubber, which is held in position by winding a fine wire around the flange.

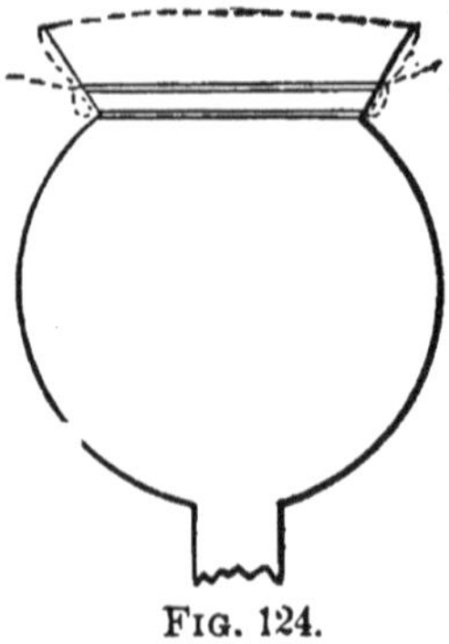

Fig. 124.

The rubber extending beyond the wire is then turned up and wired again, just above the first wiring. This usually gives a water-proof joint. The superfluous rubber is then cut off.

Apparatus for Specific Gravity by Balancing.—The simplest form of apparatus consists of two glass tubes, about 1 yard long and ¼ or ⅜ of an inch internal diameter, connected by a piece of rubber tubing about 10 inches long, the joints being wired and the whole held in a vertical position by clamps, or on a support. The two tubes may advantageously differ considerably in size, as illustrated in Fig. 125, where the larger is about ¾ in. internal diameter. Its lower end carries a perforated cork, pierced by a smaller glass tube which may either connect with a rubber tube, or be bent twice so as to form the other upright arm. This form needs no funnel. The ends of the large tube should be rounded in the flame before inserting the cork, which it is well to soak in paraffine.

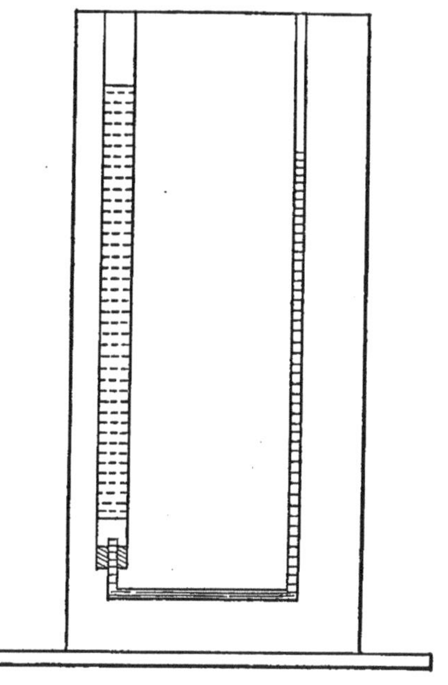

FIG. 125.

Apparatus for Atmospheric Pressure.—The barometer-tube is 3 ft. long. The device for removing air-bubbles is a feather about 2 in. long, cut down to ½ in. in width, fastened firmly by thread to the end of a straight, stiff iron wire 3½ ft. long.

Apparatus for Specific Gravity by Barometric Columns. (See Fig. 67.)—A wide-mouthed bottle is fitted with a cork carrying three tubes, the two longer being about 2 ft. long. The third is bent at right angles a short distance below the cork, and has attached to it 6 or 8 in. of rubber tubing, which terminates in a mouth-piece formed from glass tubing 3 or 4 in. long. There is also provided a plug of glass rod with which to replace the mouth-piece when the tube has to be closed. In order that the bottle may be air-tight, the cork with the three tubes inserted is immersed for some moments in melted wax. It is then quickly removed from the wax, the tubes cleared by blowing through them, and the cork crowded into the bottle as tightly as possible.

HEAT.

Apparatus for Testing Thermometers.—A piece of large glass tubing about 1 ft. long is supported vertically. In the upper end is a cork with two holes, one for the thermometer and one for the steam-pipe. In the lower end is a cork perforated for the escape of the steam.

Apparatus for the Latent Heat of Steam. (See Fig. 75.)—*D* is a glass bottle of 2 to 4 l. capacity with the bottom cut off, or a tin cylinder with a hole in the bottom fitted to carry a cork. Make the condensing coil by bending four or five feet of the common size of glass tubing into a spiral, with turns about 1 in. apart. The diameter should be such that when placed in *D* there will be about an inch between the tubing and the sides of the vessel; the height of the coiled portion of the tubing should be about three fourths that of *D*. The curved form required may be obtained by bending the tube a few degrees at very frequent intervals. Sharp bends weaken the apparatus. Pass the lower end of the tubing through the cork inserted in the bottom of *D*, and cut it off an inch below. The whole arrangement should be supported at such a height that a beaker may be placed beneath. The upper part is supported by string or wire. The steam-pipe is in three pieces, united by rubber "connections." Make the steam-trap of 3-in. tubing, the entrance tube passing just through the cork, the exit projecting in about an inch. Wind woollen cloth or rags tightly around steam-pipe, and trap to within ½ in. of the level to which *D* is to be filled with water. Secure the wrapping by cotton strings. Arrange the covering so that the pipe may be easily disconnected at *H*. The weight of *D* with cork and coil may be determined, and the value marked on the apparatus.

Apparatus for the Expansion of Gas.—I. (See Fig. 79.) Vessel *B* and support as in the apparatus for latent heat. *A* is a round-bottomed chemical flask of 250 to 300 cu. cm. capacity, with the neck drawn down to the size of the tube *DC*. A good substitute is a glass reagent bottle, with a paraffined cork perforated for *DC*. The tube, about 3 ft. long, is in two pieces, the connector being 3 or 4 in. from the bend. It projects about 2 in. above the cork in the bottom of *B*, and communicates with *A* by a rubber connector, or, if a bottle be used, by passing through the cork. Fill *A* with dry air * (or any desired gas), and attach it to the end of *DC*. If *A* is a bottle, the tube must make an air-tight joint with the cork; if a flask, the joints of the connector must be wired. Disconnect *DC* at *D* and draw a globule of mercury into the long part of the tube so as to form an index about ¼ in. long. Fill *B* with ice-water, let it stand 5 min., then, by gently tipping the tube, bring the index about 4 in. from one end, and attach this end to the connector *D*. If the gas continues to contract (as shown by the movement of the index), disconnect, move the index back, and connect again. Finally, the index should remain at rest a few inches from *D*. Support the meter-stick *E* on two blocks for ease in adjustment, lay *DC* upon it, and hold them together by rubber bands.

II. (See Fig. 80.) The jacket is a large glass or tin tube about 12 in. long, fastened to a base-board by strips of leather. A piece of small glass tubing is cut off, about 18 in. long. One end is melted so as to nearly close it; then it is warmed, and air as dry as possible drawn through it for a few minutes so as to expel any moisture. A globule of mercury is then drawn into the open end so as to form a column about 0.5 cm. long. This index is slid along the tube till about 7 in. from the partially closed end, which is then melted completely together, thus imprisoning a volume of air between the index and the closed end of the tube. The other arrangements are similar to those of Exercise 9.

Apparatus for the Expansion of a Liquid.—Test-tube with perforated cork, carrying a small-sized glass tube about 18 in. long. If the test-tube used is so long that it cannot be completely covered with the ice-water, a portion of the top may be cut off and the edges rounded in the flame. The length on the tube,

* Arrange a large bottle with a long funnel-tube and a delivery-tube, so that when water is poured into the funnel air is driven out through the delivery-tube. Arrange two or three wash-bottles containing strong sulphuric acid so that the air will be dried by the acid as it is driven from the bottle. From the last wash-bottle lead a tube inside the vessel to be used for *A*. By pouring water into the bottle the vessel may be filled with dry air. It is then connected with the apparatus, the opening being kept closed as much as possible during the operation. For any other gas substitute a generator for the air-bottle.

corresponding to 1 cu. cm., is determined as explained in Exercise 4, Mensuration. The glass tubing is generally of very uniform bore, and the average value as determined by two or three trials on one piece, would prebably be sufficiently accurate for all. If desired, the determination of this value might be made a part of the exercise. The specific gravity of the alcohol at the average temperature of the room is determined and marked on the bottle. A small flask may be used instead of a test-tube.

Apparatus for the Expansion of a Solid.—*Second Method.* (See Fig. 78.) The rod *R* is round, of brass, about 24 in. long and ¼ in. in diameter. The jacket is tin pipe or glass tubing, about ¾ in. internal diameter and ½ in. shorter than the rod. The ends are closed by double-perforated corks carrying the rod and the steam-tubes. One end of the rod screws into a brass plate *B*, which is screwed to the block *A*. The other end is in line with the brass rod *D*, which has a thread, about 40 to the inch, cut on its inner end for about 1 in. This screw works in a brass plate *E*, fastened to the block *F*. The rod *D* passes through a hole bored in *F*, in which it can just turn, and comes out in the centre of the clock-dial *G*, which is attached to the other side. It carries a pointer, *H*. The end of the jacket nearest *F* is held firmly by the support *C*. The corks are cut thin, and the steam-tubes are sunk into the cork and the edge of the jacket, as shown, the cork coming close against *B*. At the other end the rod is level with the outer edge of the cork. In order to show contact between *D* and the end of the rod, connections are made so that when *D* touches the rod an electric circuit is completed. The author uses the Edison current through a 16 c. p. lamp. A telegraph-sounder, galvanometer, electric bell, etc., would answer equally well.

Apparatus for the Expansion of a Solid.—*First Method.* (See Fig. 77.) The rod *C* is of brass, 24 in. long, ¼ in. diameter. Steam-jacket and connections as in Fig. 78, just described, with the addition of a short piece of tin tubing set into the middle of the jacket, its upper end closed by a cork perforated to carry a thermometer. The base *B* is a board about 30 × 6 in., thick enough to prevent warping. To this a heavy rectangular block of wood, *D*, is firmly attached as shown. The rod passes through the corks, and projects about ¼ in. from each end of the jacket. One end is firmly set into a hole ¼ in. deep, bored in *D*, at such a height above the base-board as to allow the rod and jacket to slightly incline towards the other end. The pointer *f* consists of a piece of square brass rod. Its lower end is pivoted in a mortise cut in a block of wood, *g*, which is firmly fastened to the base-board. A little above, it is also attached by a pivot to the end of the rod at *c*, thus forming a second-class lever. To the upper end of this rod a fine needle is attached by whipping it on with waxed thread and then covering the whipping with shellac. A millimeter-scale, *S*, is attached to an upright rising from the side of the base-board and steadied if necessary by two wire guys, *y*, *y*. The pointer should magnify at least twenty times. Through holes in the corks at the ends of the steam-jacket are passed glass tubes, *i*, *i*. The tube at the higher end of the jacket is connected by rubber tubing with a flask, *F*, holding about .5 litres, used to furnish steam. One flask will serve for several pieces of apparatus.

DYNAMICS.

Apparatus for Exercise 1.—Set a strong screw-eye firmly into the ceiling, or other suitable support, at least 6 ft. above the table—the farther above it the better. From this suspend by a strong iron wire a 10 or 12 lb. weight, or a box (cigar, starch, etc.) loaded to about that weight. Care must be used that the wire and eye are of sufficient strength. Two or three feet of common cotton string are attached to the weight.

Friction Apparatus.—The board is that used in the inclined plane (Fig. 89). The blocks may be sawed from hard-pine floor-board. Instead of an 8-oz. balance a rubber strip may be used. A light strip of wood has a scale on its lower half.* A strip of rubber about 18 in. long and $\frac{1}{4} \times \frac{1}{8}$ in. is fastened to the upper end, as in the apparatus for elasticity (Fig. 95). One or two marks on the strip serve as references. A rod attached to the lower end of the strip, as in Fig. 95, holds the connecting cord.

Apparatus for Composition of Angular Forces.—Near one edge of the table drive two nails about $2\frac{1}{2}$ ft. apart; hook over these the rings of two of the balance and connect them by a piece of cord of such length that when gently pulled at the centre the

---

* This may be a centimeter scale marked on the wood, or the apparatus may be graduated in grams by testing with weights.

balances do not lie in a straight line. Take a second piece of string, about 2 ft. long, and tie its centre around the first piece of string so that it will slide upon it, but the knot will not slip. To the ends of this piece of string attach two other balances.

Apparatus for Parallel Forces. (See Fig. 88.)—A 25-lb. balance is suspended from the ceiling, or some other support, by an iron wire. A hole is bored through a meter-stick at the 50 cm. mark, and the meter-stick is suspended from the hook of the balance by a loop of strong cord passing through the hole. Two pieces of cord about 6 inches long, having loops at their upper ends to slide on the stick, are attached to the rings of two 25-lb. balances. The meter-stick may be mounted in some such way as that in Exercise 12, Dynamics.

The Inclined Plane. (See Fig. 89.)—Two pieces of matched hard pine floor-board may be used to form a board 7 in. wide; they may be held together by two cleats on the under side. The length is about 6 ft., and the boards are planed. To prevent slipping, the lower end may rest against a couple of nails or a cleat fastened to the table. The general form of support is indicated in the figure. In another form, which is easily constructed, a piece of hard-pine floor-board about $2\frac{1}{2}$ ft. long is nailed to to each end of a starch-box weighted with stones or sand. The tops of the uprights are connected by a cross-piece nailed on, and several holes are bored in opposite pairs in each upright to receive the supporting rod.

A cheap form of roller-skate, in which the axles are rigid and do not turn, and the wheels as far apart as possible, makes a good carriage.* The straps are removed, the axles oiled, and to one end is attached a screw-eye, to which is fastened a piece of fish-line about 12 in. long, with a loop at the end. The straps from these skates can be used to attach glass tubing to boards. If no skate can be obtained, almost any small carriage can be used—a tin cart for instance, provided the friction is not very great and the wheels run smoothly. As the friction is to be corrected for, the amount, within reasonable limits, is not important, provided it is constant. Another substitute

---

* The wheels on these skates are often set so that the skate tends to run in a curve, and they may need adjusting before they are used for the first time.

might be a piece of board, or a box, about 4 × 6 inches, mounted on easy-running casters. The total weight of the carriage when loaded should be about 15 lbs. If a skate be used, a cigar or chalk box, filled with pieces of iron, lead, etc., may be tied on to it, or fastened by screws. In loading the carriage care must be used to get the centre of gravity as low and as near the centre as possible, so that the arrangement will not readily capsize if accidentally pulled sideways. This experiment calls for two students. It could be modified for one student by leading the cord over a pulley at the upper end of the board and attaching a scale-pan of known weight, the force required being measured by weights placed in the scale-pan. In this case the carriage, etc., need not weigh over 4 or 5 lbs. For Part II a piece of hard pine $\frac{7}{8}$ in. square and 8 in. long, having a hole bored in each end, is screwed crosswise to the front end of the skate. A similar piece has at its centre a screw-eye or loop of string to hold the balance. The two are connected by pieces of strong cord about 4 ft. long.

Collision Apparatus.—*First Method.* (See Fig. 92.) The base-board is about 3 ft. long and 5 or 6 in. wide. Fasten a meter-stick on edge in the centre of the board parallel to its length. At diametrically opposite points in each ball drill holes 1 or 2 mm. in diameter and 2 or 3 mm. deep. Prepare wooden plugs which fit tightly in the holes, projecting 2 or 3 mm. beyond them. Place the ends of the suspending cords or wires in these holes and force the plugs firmly in. The balls may be readily detached by withdrawing the plugs. Suspend the balls at the point of a V formed by the two wires, arranged, as shown in the figure, so that the balls swing freely in one vertical plane. This may be done by attaching to the ends of a board about 1 × 2 ft. two screw-eyes, and passing the wire from the ball through one eye across the board through the other eye, and back to the ball. The board is fastened to the ceiling. The second ball is suspended like the first, the distances between the screw-eyes being so arranged that when both balls hang at rest they just touch, and their centres are in the same line, parallel to and just above the meter-stick. The two electro-magnets are supported above the meter-stick, so that they may be moved either vertically or horizontally in the plane in which the balls swing. Find by trial the point on the balls which touches the ends of the magnets. At each of these points drill a small hole, plug it with wood, and drive a small iron tack into the wooden plug. If the magnets are of sufficient strength the balls can be held in any position desired, and released by breaking the circuit. By placing both magnets in series, the balls may be released simultaneously. Have four rectangular blocks of wood with two upright cards tacked on them on opposite sides. Place these blocks beside the meter-stick just clear of the balls, and move them along until on sighting across their sides the line of sight strikes the centre of the ball. The number on the meter-stick struck by the line of sight is the reading for the ball. It saves time to mark the weights of the balls on them. The author uses the magnets out of old telephone bells, and the Edison current through a 16 c. p. lamp. With these coils and this current there is some danger of overheating if the circuit is kept closed too long at a time. The electro-magnets may be supported on a block of wood which slides on the meter-stick, and carries an upright rod thrust through a large cork, into which is also thrust one of the projecting ends of the magnet. The cork may be slipped up and down the rod, and the block wedged in any position on the meter-stick. The most convenient support for the magnets, however, is a wooden clamp. The balls weigh about 10 and 30 grams, respectively. If putty be used, a weighed quantity may be spread over one ball, and its weight added to that of the ball.

Second Method. (See Fig. 93.) Remove the handles of two pint tin pails and suspend them from the ceiling by wires, whose upper ends are about two feet apart. The lower ends hook into the holes in the pails where the handles were attached. Make a spiral spring about three inches long by twisting spring-brass wire tightly around a broomstick. Attach this spring to one of the pails at a point 90° from the suspending wires. Arrange the distances between the points of suspension, so that when the spring is compressed the pails will hang side by side with the spring between them. Beneath the pails place two meter-sticks end to end, about 1 cm. below, scale side up, and in the plane in which the pails swing.

Apparatus for Boyle's Law. (See Fig. 96.)—The apparatus is the usual form, and may be obtained of instrument-makers. The tube is made of a piece of "American glass" tubing of about 8 mm. internal diameter and about 130 cm. long. One end is sealed up in the usual way, care being used to make as square an end as possible. The tube is bent into the form shown in the figure, the horizontal portion being 5 or 6 cm. long, and the short arm about 40 cm. long. The tube is fastened to an upright by strips of leather. A meter-stick is attached to the board, parallel with the long arm. If the end of the short arm is properly sealed, a 40-cm. scale cut from a meter-stick may be attached parallel to the short arm, and the experiment carried on in the usual way. The writer prefers, however, to calibrate the tube, and hence measure directly, the volume of the gas. A convenient method of calibrating is to fill the short arm about half full of mercury, invert it, get it as free from air-bubbles as possible, and read the position on the 40-cm. scale corresponding to the highest point of the meniscus.* The mercury is then weighed, and the volume obtained in the usual manner. Two or three more determinations are made with larger quantities of mercury, carrying the measurements nearer to the bend. From these determinations a calibration-card is made out, giving the readings on the centimeter scale for each cubic centimeter, and the students are shown how to find intermediate values by interpolation. It is not necessary to carry the calibration-card above the lower half of the short arm. A scale may be spaced off upon card-board, starting from about the middle of the tube and running down to the bottom, giving the volumes directly in cubic centimeters.

The arrangement for removing air consists of a stout iron wire, to one end of which is fastened by waxed thread a strong feather (from a feather-duster), cut down to about $\frac{1}{8}$ in. in width and 1 in. in length. The wire is long enough to reach from the top of the tube to the bend, and the upper end is twisted into a ring so that when not in use it may be hung on a nail driven into the side of the board near the top. The arrangement for reading consists of a piece of stiff card-board tacked on to a piece of a meter-stick about 10 cm. long. The support for this exercise is also used for the exercise on Specific Gravity by Balancing. A board about 3 ft. long, 8 in. wide, and $\frac{7}{8}$ or 1 in. thick, is sup-

FIG. 126.

* For correction for meniscus see S. and G., vol. 1, p. 109.

ported vertically, either by nailing its lower end on to a weighted box, or by fastening it to a base-board about 1 ft. square. The Boyle's-law apparatus may be secured to one side and the Specific-gravity apparatus to the other.

Exercises 11 and 12.—The meter-stick is supported by means of a double hook of heavy wire (as shown in Fig. 126), which holds a wire nail passing through a hole bored in the 50-cm. mark. The spring may be constructed of spring-brass wire wound around a broomstick.

Apparatus for Elasticity of a Solid. (See Fig. 96.)—The block *A* is screwed to some firm upright, either the side of the room or the upright used for Boyle's Law apparatus. 96 cm. below it attach the second block in the same manner. The meter-stick is secured to these two blocks as shown. The block *B* is secured to the block *A* with two long wood-screws, not shown in the figure. The reading-card *C* may be made by tacking a card to a 10-cm. scale or similar piece of wood, and is held in position by elastic bands, *R R*, slipped onto the meter-stick before it was attached to the blocks. One end of each rubber strip is turned up around a piece of lead-pencil or penholder about 2 in. long, *D* in diagram, and is secured by paper-fasteners or rubber cement. The rubber should not be stretched around *D* so tightly as to prevent the easy withdrawal of *D* if one scale-pan is to be used for both strips. The scale-pan may be a piece of cigar-box wood of sufficient size (say 4 × 4 in.) to hang clear of the upright when the strip is attached to *A*. It is suspended from *D* by light wire. To attach the strip, loosen the screws holding *B*, pass the end of the strip up between *B* and *A*, so that about 1 cm. projects above, and tighten the screws again. Find by trial a point on the strip corresponding to a length of about 90 cm. when the greatest weight that is to be used is in the scale-pan. Mark this point with ink (it may be well to letter it *b*); half-way between this point and the lower side of *B* mark a second point, *a*. Prepare the second strip in the same way. One scale-pan can be used for both, or they may be provided with separate pans.

LIGHT.

Apparatus for Focal Length. (See Fig. 99.)—The meter-stick is secured to a suitable base-board by wire nails or screws passing through holes bored in the meter-stick, the heads being countersunk. This does not injure the meter-stick for other use. The blocks are about 2 × 2 × 3 inches, grooved to slide on the meter-stick with a little friction, the bottom of the block resting upon the base-board. The screen *S* is of cigar-box wood, about 5 × 6 inches, covered on one side with card-board or white paper, and is attached to the right-hand end of the block. When the block is placed upon the meter-stick the screen should be at right-angles to the meter-stick and perpendicular to the base-board. The same precautions should be used for the other uprights. Another piece of cigar-box wood similarly arranged is fitted to hold the lens. A spectacle lens of about 6 inches focal length answers the purpose. The chimney is supported on three corks, which may be glued to the block and be grooved on the upper surfaces to receive the edges of the chimney. To use as a photometer (See Fig. 100), prepare a support similar to that for the lens, cutting an opening in it about 2 × 2 inches. Over this opening fasten a piece of unsized paper (that entirely free from gloss) with a paraffine spot about 1 in. in diameter in the centre. To make paraffine spot, cut in a piece of stiff paper a circular hole about 1 inch in diameter, lay this paper upon the unsized paper, rub the portion exposed with the end of a paraffine candle, then warm the paper (with a hot iron or by other means) enough to melt the paraffine. The block *C* might be provided with a strip of cigar-box wood fastened at right angles to the length of the meter-stick so as to carry more candles in line.

For Ex. 3 the instrument is provided with a screen about 5 × 5 inches, mounted as in the other exercise, and having in its centre an aperture 1 cm. square. For this exercise the light should be as near to a point as possible; hence in front of it should be placed a card-board screen provided with an opening about 1 cm. square, and held by a clamp or suitable support. Or the light may be placed in a small box, open at the top, and having a small aperture cut in one side at the level of the flame. In preparing the apparatus for use in Exercises 1, 2, and 4, candles should be provided of such length that at the beginning of the exercise the centres of the flames will be slightly above the horizontal line passing through the centre of the lens, the grease spot, or the opening. It will save time in correcting for true position to make the blocks an even metric length, say 8 cm., and mark this length upon them. Blocks to carry candles should have a line ruled across the top midway between the ends and at right angles to the length, to mark the centres of the candles.

Rumford Photometer. (See Fig. 101.)—The meter-sticks, blocks, etc., arranged as above. The angle between the meter-sticks should bring the two shadows about half an inch apart. The screen, about 5 × 5 in., attached to the base-board by a block 2 × 2 × 5 in. The rod may be a pencil or pen-holder set vertically into a hole in the base-board. Any source of light may be used—different-sized candles, small kerosene lamp, incandescent lamp, etc. The screen *SS'* may be of stiff card-board, about 1 × 2½ ft. The most suitable light for Exercises 1 and 4 is an Edison 16 c. p. lamp. In Ex. 1 the image of the filament is very easy to focus, and the electric lamps are to be recommended on the score of cleanliness. They may be mounted in a single pole cut-out attached to the apparatus, from which slack connecting wires may be led. The plane of the filament should be at right angles to the screen. It is well to provide some means of screening the eye from the light on *L* while comparing shadows. The distance from the rod to the screen might be measured along the line of the meter-sticks and marked on the apparatus.

SOUND.

Sonometer. (See Fig. 102.)—Strong fish-line may be used for the cord *cc*. The nails for securing the string should be from 2 to 3 inches apart. Those having sharp corners should be avoided. The wires should be of two different sizes,—say Nos. 28 and 30, spring-brass,—and should be secured both to nails and to balance hooks by taking a round turn, then twisting the end of the wire around the main portion. The triangular blocks may be 1 inch high.

APPENDIX D.

NOTES AND REFERENCES.

For the convenience of teachers, and in order to show how to connect this course of exercises with the usual class-work, there follow detailed references to five text-books, whose full titles and publishers are as follows:

Avery's *First Principles of Natural Philosophy.* New York, Sheldon & Co.

Avery's *Elements of Natural Philosophy.* New York, Sheldon & Co.

Gage's *Elements of Physics.* Boston, Ginn & Co.

Gage's *Introduction to Physical Science.* Boston, Ginn & Co.

Hall and Bergen's *Text-book of Physics.* New York, Henry Holt & Co.

It is thought that one of these will resemble any book likely to be used closely enough to render the making of a parallel series of references an easy matter. The numbers refer to sections, unless otherwise indicated.

In order to economize space, all of the notes are given a telegraphic brevity of form. The title of each subject is immediately followed by some suggestion of the particular educational ends which the succeeding exercises are intended to serve. These notes will in some sort account for the order in which the subjects are placed.

In the hints to teachers the degree of accuracy which may be expected is given approximately. Where the result would vary with each pupil, an example drawn from actual work is substituted for a numerical statement of the probable percentage of error. Occasional reference for the use of the teacher is made to the books mentioned at the end of Appendix A.

MAGNETISM.

How to plan and conduct an experiment, apply the knowledge obtained, take methodical notes, and draw inferences. Study of phenomena due to condition ; inductive reasoning. This work should be made the basis of considerable class discussion, criticism of notes, etc.

Exercise 1.—Show how to tabulate results neatly ; advantages of this method of recording results. Rudimentary ideas of work and energy might be brought out.

REFERENCES. Avery's *Principles*, 1–35, 47, 283, 284 ; *Elements*, 1–9, 13–21, 64, 304, 305. Gage's *Introduction*, 1–4, 11, 167 ; *Elements*, 1, 2, 12, 91, 188 (paragraph 1). Hall and Bergen, 1–4, 26.

Exercise 2.—In Fig. 2 the magnet should be represented as horizontal. This arrangement is preferable to the compass, because no new apparatus is introduced.

REFERENCES. Avery's *Principles*, 285, 288 ; *Elements*, 310, 317 ; Gage's *Elements*, 197.

Exercise 3.—REFERENCES. Avery's *Principles*, 286 ; *Elements*, 306. Gage's *Introduction*, 165, 166 ; *Elements*, 186, 187. Hall and Bergen, 290–292.

Exercise 4.—Try effect of direction of stroke. A No. 15 needle is good. Iron dust may be used for Exp. 6. Can pupils explain results of Ex. 1 from the results of this Exercise?

REFERENCES. Avery's *Principles*, 280–282, 287, 291–293 ; *Elements*, 302, 303, 307, 311, 320. Gage's *Introduction*, 168, 169 ; *Elements*, 185, 194, 196, 198. Hall and Bergen, 286, 288.

Exercise 6.—Different methods might be given to different students, and comparative results subsequently discussed. Draw attention to the fact that the diagrams obtained are sections of the field of force. Models of wood might be easily made, showing the field in three dimensions, using wires for lines.

REFERENCES. Avery's *Principles*, 204, 289, 290, 294–297 ; *Elements*, 313–319. Gage's *Introduction*, 173–176 ; *Elements*, 191–193. Hall and Bergen, 287, 289, 293, 295.

ELECTRICITY.

In general, these exercises illustrate the conditions under which phenomena develop and those affecting the degree of development, elementary physical concepts, use of instruments in investigations and determination of comparative values, the diagrammic method of recording results, and the method of drawing inferences from comparison of a number of values.

Exercise 1.—Experiment 1 might be performed by the teacher before the class. Part of the class might try placing the wire over the needle.

REFERENCES. Hall and Bergen, 303.

Exercise 2.—It would be well to leave strips of copper and amalgamated zinc in dilute sulphuric acid on a closed circuit for a day or two, to emphasize the diminution of the zinc and permanency of the copper. If possible, let the pupils themselves discover that the power to deflect the needle only exists while zinc is consumed. By leaving the conductor over a compass, effects of polarization might be noted. If various commercial cells are available, let the students examine them, and point out how each fulfils the conditions observed to be necessary. For pictures of many modern cells, see the advertising pages of the *Electrical World*.

REFERENCES. Avery's *Principles*, 246–249, 254–263; *Elements*, 373–377, 381–386. Gage's *Introduction*, 128, 130, 135, 136, 138–141; *Elements*, 151–154, 158–166. Hall and Bergen, 305–307, 310.

Exercise 3.—In case of limited equipment, two or three large cells with long conducting wires might be used for a half-section, and this exercise alternated with Exercise 4. In this connection the method of recording results by diagrams, with arrows to indicate motion, might be suggested. Results may be recorded very rapidly if a set of diagrams without the arrows have been previously placed in the note-book. Reading by reversal is the best method. A reverser is to be recommended as a convenience; but, since it is not absolutely required, it is not referred to in the instructions, nor does it appear in the diagrams, although given in the lists of apparatus in Appendix B. Should it be used, the instructor should show the class the manner of connecting it and its action. (See Appendix C.) It might be permanently attached to the galvanometer, the two free posts being considered as the galvanometer-posts, the galvanometer sign in diagrams covering both.

REFERENCES. Avery's *Principles*, 277; *Elements*, 390, 391. Gage's *Introduction*, 131, 132, 149, 150; *Elements*, 156, 157, 172–175. Hall and Bergen, 313, 314.

Exercise 4.—The effect of temperature on resistance may either be given or be illustrated by experiment. Iron-spiral, sensitive galvanometer, burner, etc. If the class have no idea of the yard, a yard-stick might be shown to give some idea of the lengths of wires used.

REFERENCES. Avery's *Principles*, 216, 220, 270, 271 ; *Elements*, 378, 397–402. Gage's *Introduction*, 143, 144, 152, 191, 192, 197 ; *Elements*, 169, 170, 177, 178, 237, 238, 246.

Exercise 5.—Fill cells well up to reduce the internal resistance. It might be well to explain what is meant by "cross-section," and how it is indicated by number. For tables of gauge numbers, see Hall and Bergen, p. 372 ; Chute, p. 365.

REFERENCES. Avery's *Principles*, 268, 269, 312, 313 ; *Elements*, 387–389. Gage's *Introduction*, 142, 159, 187–190 ; *Elements*, 167, 168, 232–236.

Exercise 6.—Internal resistance should be made high by using only enough fluid to immerse the plates to about one third their depth. The use of connectors avoids twisting the ends of wires. The diagramic sign in Fig. 27 is simplified in Figs. 29, 30, and 31, but is essentially the same. Either form is allowable. In connection with the ohm, a piece of wire about 1 sq. mm. c. s. would give an idea of the size of the mercury column. The explanation in the instructions is given chiefly with the idea of showing how the standard is obtained by attaching specific values to all the variables of which resistance is a function. An ohm of some conductor might be shown, say about 4 meters No. 28 copper wire. For description of B. A. ohm coil see Stewart and Gee, vol. ii. p. 161 ; *Electrical World*, Jan. 10, 1891. Various forms of resistance boxes are described in Stewart and Gee, pp. 133, 148, 158.

REFERENCES. Avery's *Principles*, 250, 264–267 ; *Elements*, 379, 380. Gage's *Introduction*, 153–155, 160–163 ; *Elements*, 177 (paragraph 2), 179, 183. Hall and Bergen, 316, 319.

Exercises 7, 8A, 8B.—Alternative methods : 7, where no resistance-boxes are available ; 8A and 8B, where some form can be used. For relative resistance iron may be roughly taken as six and German silver as twelve times copper under the same conditions. For tables, see Chute, p. 364 ; Hall and Bergen, p. 374. In 8A and 8B, bodies suited to the capacity of the box would be required. Lengths of various wires stretched back and forth between tacks on a board, lengths on racks for Exercise 7, etc., or even the wire coils could be used. Wires wound in coils would not give very accurate results, owing to inductive effects. In 8B, two or three sulphate-of-copper cells in series would give a good current to work with. Values could best be checked by results obtained by the instructor on the same apparatus. For full discussion and description of these methods, see Stewart and Gee, vol. ii. pp. 94–97; 105–110. Daniell, p. 591.

References. Gage's *Introduction*, 156.

Exercise 9.—This might be given directly after Exercise 5, or explanation of the results of Exercise 6 deferred until E. M. F. had been discussed. Dilute sulphuric acid might be used. There may be less danger of getting acid on hands or clothes if vessels of water are provided in which plates can be rinsed before handling. For discussion of electrical units, see Everett, pp. 140–148. For standard cells, *Electrical World*, June 20, 1891. Methods of measuring E. M. F., Stewart and Gee, pp. 237–249; Kohlrausch, pp. 222–225.

References. Avery's *Principles*, 218, 219, 251–253, 273, 274; *Elements*, 378. Gage's *Introduction*, 133, 134, 146, 157, 158, 186; *Elements*, 155, 180–182, 184. Hall and Bergen, 311, 317.

Exercise 10.—Results of Experiment 1 may be well recorded by diagrams and arrows. In Experiment 3, the current strength might be changed by using mercury-cups and coils as in Exercise 6, or the apparatus for Exercise 7. Suggestion: Let pupils devise some method utilizing apparatus that they have already used.

References. Avery's *Principles*, 202, 203, 205, 275, 276, 298–300; *Elements*, 392–396. Gage's *Introduction*, 170, 171, 193–195; *Elements*, 171, 188 (paragraph 2)–190, 240–244.

Exercise 11.—This exercise deals with the fundamental principles of the dynamo and the motor. It may be worked by two students, one to watch the galvanometer and the other to handle the magnets. If desired, however, it may be worked by the teacher and class together. In that case, some sort of an indicator had better be attached to the galvanometer needle, such as that described in App. *C*. If possible it should be followed by instruction on the motor and dynamo, which might be illustrated first by a model, as in App. *C*, and then by a real motor which can be taken apart. A talk on the commercial applications of dynamic electricity would be interesting. For illustrative cuts, see *Electrical World*, *Scientific American*, etc. For explanation of action of dynamo, see "Shaw" or "Experimental Science."

References. Avery's *Principles*, 206, 278, 303–311, 314, 315; *Elements*, 403–414. Gage's *Introduction*, 178–185, 196, 198, 199; *Elements*, 200–206, 245.

For practical applications of electricity, see such publications as the *Electrical World* and the *Scientific American*. The following may be of interest: Course of Electrical Reading, Dr. Lewis Bell, *Electrical World*, Aug. 8, 1891; Hertze's Experiments, *Electrical World*, July 25, 1891; Electrical Units, *E. W.*, Jan. 3, 1891.

MENSURATION.

Determination of single values in terms of various standard units. Training in the use of measuring instruments. Determinations requiring simple mathematical work in finding required results. Results calculated from experimental data. General methods of quantitive work. Errors. Determination of one value by several special methods.

Exercise 1.—Practice in the use of linear scales. Fair values, 39.45 to 39.30, crosses about 40 cm. apart. As this is the pupil's first quantitative work, full tables are given. The less the distance between the crosses the greater the error. For accurate determination of length, see Stewart and Gee, vol. i. pp. 1–45.

References. Avery's *Principles*, App. B; *Elements*, 22–30, 33–36. Gage's *Introduction*, App. Sect. A; *Elements*, App. Sect. A. Hall and Bergen, App. 1.

Exercise 2.—Average of trials should be very near 3.14. Examples: 3.14, 3.14, 3.14, 3.15, both methods, two circles. For accurate determination of volume, see Stewart and Gee, pp. 104–113. Area, pp. 95–104. Calibration, Pickering, pp. 37–39; Stewart and Gee, vol. i. 7, 109. Cleaning mercury, Pickering, p. 35.

Exercise 3.—Method B should give results slightly higher than A or C, which should give results about alike. The difference may be 0.2 cu. cm. Examples: A 7.5 cu. cm., B 8.0 cu. cm., C 7.4 cu. cm. Chief error, careless reading.

References. Avery's *Elements*, 31.

Exercise 4.—In tubes 10–12 mm. diameter, calculated and measured results should agree to first decimal place. Two methods should agree closely. Examples: .626 sq. cm., .627 sq. cm., by one student; .625, .629, by another. For description of accurate determination of weight, etc., see Stewart and Gee, vol i. pp. 61–94. Pickering, pp. 46, 47.

REFERENCES. Gage's *Introduction*, 12, Hall and Bergen, 5, 6.

Exercise 5.—Practice in the use of balances and scales. Chief error in reading spring balance. Bodies weighing 50 to 75 grm. best. Examples of results : 28.29 g., 28.3 g., 28.08 g.

Exercise 6.—Given if needed. For full discussion of errors, etc., see Kohlrausch, pp. 1–23 ; Stewart and Gee, vol. i. App. A.

Exercise 7.—An exercise in weighing liquids. For accuracy, a correction should be made for the liquid adhering to the sides of the vessel from which liquid is poured when the two are mixed. Probable error, 0.3 g.

DENSITY AND SPECIFIC GRAVITY.

Determination of physical ratios by measurements of two independent values. Special methods based on mathematical use of knowledge already obtained. Interpolation of experimental data in formulæ. Indirect measurements.

Exercise 1.—Object, to give a clear idea of density and its relation to specific gravity. For general discussion and accurate methods of determining relative density, etc., see Stewart and Gee, vol. i. pp. 114–162.

REFERENCES. Avery's *Principles*, 165–167 ; *Elements*, 241–243. Gage's *Introduction*, 8, 53, 54 ; *Elements*, 7, 62, 63. Hall and Bergen, 47, 48, 52, 53.

Exercise 2.—The method of determining specific gravity by the specific-gravity bottle is given first because this determination calls for no indirect measurements, being the direct determination of the weights of equal bulks ; hence the principle is more easily grasped by the student than the specific-gravity determination involving indirect measurements. Probable error, 2 per cent ; example : copper-sulphate solution, 1.05 and 1.07, by different pupils with different apparatus.

Exercise 3.—A necessary preliminary to the determinations following. The student is expected to discover Archimedes' principle as a fact, the explanation being taken up later in connection with the experiment on Liquid Pressure. It is very essential that this fact should be thoroughly grasped before taking up the succeeding exercises. If the fact that a body loses weight when immersed in a liquid is not known to all the class, a simple qualitative experiment will readily show it. It is best to use a body of low specific gravity (1.5 to 2), 10 to 20 g. weight. Example: Electric-light carbon—loss of weight, 8.03 g.; wt. of water displaced, 8.05 g.

REFERENCES. Avery's *Principles*, 162; *Elements*, 237, 238. Gage's *Introduction*, 51, 52; *Elements*, 61. Hall and Bergen, 50, 51.

Exercise 4.—Good examples of indirect measurement. Experiment 1: For best results the body should weigh 30 to 50 grams. The best thing for suspension is a piece of the finest copper wire obtainable,—say No. 30. It may be well to explain the error due to the weight of the wire, and the reasons why it may be neglected. Example: Carbon, 1.7, 1.5; different pupils with different pieces of apparatus and different pieces of carbon. Exp. 2 optional. A cork may be used. For accuracy, it is best shellaced or dipped in melted paraffine.

REFERENCES. Avery's *Principles*, 168; *Elements*, 244, 247. Hall and Bergen, 55.

Exercise 5.—The fact that liquids exert a pressure on bodies immersed in them is usually a matter of common observation; but if necessary, this exercise may be preceded by a qualitative one demonstrating the fact. Either one of those given in the text-books, or some such experiment as the following, may serve:

Apparatus.—A lamp-chimney; sheet-rubber; glass jar large enough to hold chimney. Tie* the rubber over one end of the chimney, and thrust it into the water. The bulge of the rubber diaphragm will show that a pressure is exerted on the bottom. Other modifications will readily suggest themselves. Before taking up the laboratory exercise the class might be called upon to state the conditions that they suspect would influence liquid pressure, the conditions of working, etc. It is best to set up the apparatus so that the gauge can be under water overnight. Only general results are expected. Before commencing the exercise be sure that all the gauges are water-tight, as any leakage will make trouble.

REFERENCES. Avery's *Principles*, 39, 146–158, 171–177; *Ele-*

---

* See this exercise, App. C.

ments, 215-231, 254-267. Gage's *Introduction*, 9 (paragraph 2), 31, 45-48, 37; *Elements*, 43-45, 52-55. Hall and Bergen, 27-30, 32-38, 65, 66.

Exercise 6.—Intended not only as a specific-gravity determination, but also to impress on the student some of the prominent facts of hydrostatics. If desired, the formula $\triangle \times D = \triangle' \times D'$ could be discovered experimentally, by using liquids of known densities, and the explanation subsequently given. This exercise also explains the tendency of water to seek its level, etc., and is made the basis of instruction in all these points. The author uses both forms of apparatus, but prefers that given in App. *C*. Examples: Kerosene, 0.79, 0.79, 0.80.

REFERENCES. Avery's *Principles*, 160, 161; *Elements*, 232-234. Gage's *Introduction*, 49; *Elements*, 56-58. Hall and Bergen, 60.

Exercise 7.—The chief error lies in reading the volume with the test-tube floating in the liquid. If desired, an exercise could be introduced here on the determination of specific gravity by "flotation," using a rectangular piece of board and measuring the depth to which it sank, as compared with the thickness of the board.

Example of results: Weight of tube, 13.5 g.; wt. of water displaced, 13.5 g.; copper sulphate, 13.25 g.

REFERENCES. Avery's *Principles*, 163; *Elements*, 240, 249-252. Gage's *Introduction*, 57; *Elements*, 60, 61 (paragraph 2), 64. Hall and Bergen, 58.

Exercise 8.—Shows how the pressure of the atmosphere may be actually measured in units of weight and on given areas, and should be preceded by one showing that air has weight. Many devices to illustrate this can be obtained of dealers in apparatus,* but as they are quite expensive, one of the following may be substituted. (*a*) When an air-pump is available, a bottle holding from 1 to 2 litres is provided with a perforated stopper (either of rubber or cork which has been boiled in melted wax and inserted hot), carrying a rubber tube about 6 in. long, on which is a screw clamp.† The rubber tubing should be well wired to the glass. Connect the tubing with the air-pump, exhaust as much as possible, close the clamp tightly, and counterpoise on the scales. Open

---

* James W. Queen & Co. E. S. Ritchie & Sons.

† This form of clamp is given in E. & A.'s Cat. No. 5968.

the clamp and admit the air. With a good degree of exhaustion the increase in weight is very noticeable. An aspirator might be substituted for the air-pump.

(*b*) Where no air-pump can be obtained,* a thin glass flask, as large as is available, is used instead of the bottle, and filled with the perforated stopper, etc. For the clamp a glass plug may be substituted. Put some water into the flask and boil until the flask is full of steam. Then remove the flame and insert the cork, the tube being closed by plug or clamp. Allow it to cool, and proceed as above. Having established the fact that air has weight, apply the principles of Exercises 5 and 6, and so lead up to the principle of the barometer, considering it a case of balancing columns, where one column is air. By working in this way, familiarize the students with the principles of the barometer, and then go on to the exercise, which the teacher might prefer to work with the aid of the class, rather than to put it directly into their hands, owing to the amount of mercury required, and somewhat delicate manipulation. It may be well to make sure that the pupils understand that the object of completely filling the tube with mercury is simply to expel all air. Results probably slightly under theoretical. It might be interesting to repeat the experiment with some other liquid, and see if pupils can explain the different results.

REFERENCES.—Avery's *Principles*, 178-194 ; *Elements*, 268-281, 288-301. Gage's *Introduction*, 31, 35, 36, 40-44 ; *Elements*, 43-49, 51, 59, 60. Hall and Bergen, 39-43, 61, 62, 67-69.

Exercise 9.—A correction for capillarity might be made. Results should closely check those of Exercise 6. Example, copper sulphate solution, by Ex. 6, 1.07 ; by Ex. 9, 1.06.

REFERENCES. Avery's *Elements*, 235, 236. Gage's *Introduction*, 29, 30 ; *Elements*, 34. Hall and Bergen, 60 (paragraph 2).

For miscellaneous tables of specific gravity, see Chute, p. 361 363.

HEAT.

Graphical method of recording results ; correction for known errors ; use of more complicated formulæ ; calibration of apparatus ; more complicated calculations.

---

* A simple form of air-pump is described in "Experimental Science."

Exercise 1. Before taking up the exercise on Conduction, etc., the ordinary phenomena produced on heating a body are shown the class by means of some of the common qualitative experiments. For convenience, an outline of the apparatus used in this course is given here. It would be a matter of judgment with the teacher whether to perform them before the class, or give them out as Laboratory exercises.

Change of volume produced by a rise of temperature for solids. The apparatus given for Exercise 9, method 1, may be used. It is a good plan, however, to remove the steam-jacket and heat the rod directly. For liquids, the apparatus for Exercise 10 might be used. The liquid used is water or alcohol. If this is to be used as a lecture experiment, the liquids had best be colored by the addition of black or red ink and a strip of white cardboard placed behind the tube. The arrangement is then heated. For gases, the same apparatus as above, without the liquid. It is supported in an inverted position, the lower end of the tube being immersed in a glass cylinder containing liquid. By slightly warming the flask, a few bubbles of air will escape, and while the air in the flask is cooling to the temperature of the room, a column of the liquid will be drawn up into the tube, which will serve as an indicator. The warmth of the hand is usually sufficient to produce a marked change in the index. These arrangements are obviously models of thermometers, and may be used later to illustrate the principle of those instruments. The peculiar behavior of water as regards expansion may be illustrated with the same apparatus by placing the flask filled with water, as above, for some time in a mixture of ice and salt; upon applying heat, the liquid column will first sink and then rise. A similar piece of apparatus filled with alcohol might be used at the same time to make the contrast more marked.

REFERENCES.—Avery's *Principles*, 358, 361–364, 392–397; *Elements*, 474, 475, 483–491, 537–542. Gage's *Introduction*, 99–106; *Elements*, 102, 106, 108–113, 116. Hall and Bergen, 134–137, 189–192.

Exercise 2. Before taking up this exercise, the class should be shown the special form of thermometer used and instructed in methods of graduation. The thermometers usually need testing, the 0 point being often displaced. In actual use, the 0 correction is needed in Exercise 3 and some others. Of course, where differences are taken, no correction is needed.

This Exercise also shows the principle on which thermometers are graduated. The apparatus used in Exercise 10 may be graduated by using mercury and noting the distance that the column rises when the test-tube is first immersed in ice and then in steam. Rise corresponding to 1° F. or 1° *C.* can then be calculated. This would form a good lecture experiment in connection with instruction on the graduation of thermometers. For the temperature of steam at different pressures, see Stewart, p. 11. For general discussion of thermometers, ibid. 1-24.

REFERENCES.—Avery's *Principles*, 359, 360 ; *Elements*, 476-482, Gage's *Introduction*, 107-109 ; *Elements*, 117-123. Hall and Bergen, 138, 139, 142, 143.

Exercise 3. The effect of temperature on physical form is illustrated by heating in test-tubes such substances as wax, paraffine, water, etc. To illustrate the direct change from a solid to a gas, a few crystals of iodine may be heated in a dry flask. In connection with these experiments, explanations may be given of what is meant by the "melting-point," "boiling-point," etc. The chief object of these experiments is to place the students in possession of certain facts which form the basis of the subsequent exercises, in order that they may bring to the work fair understanding of the principles involved. Vigorous stirring and finely broken ice are the main requisites for satisfactory results. A good method of breaking ice consists in wrapping it in a stout cloth and pounding it against a brick wall. Snow may be used in place of the ice, but must be thoroughly mixed with the water and not allowed to remain in lumps. If the thermometer rises before the ice has entirely melted, it is probably due to an insufficient mixture of water and ice, or to the gas having been turned on too much.

REFERENCES.—Avery's *Principles*, 369-371, 373; *Elements*, 53-60, 495, 498-501, 508. Gage's *Introduction*, 9, 110, 113; *Elements*, 15-17, 131, 361. Hall and Bergen, 167, 168, 175, 177-184.

Exercise 4. Probably only general results will be obtained. The best vessels to use are beakers holding 15 to 25 cu. cm. If different volumes are taken by different students, the effect of quantity on the rate of fall in temperature may be observed. A comparison of the heating and cooling curves might be instructive. For law of cooling, see Stewart, pp. 24–27.

Exercise 5.—For another method, see *Journal of Analytical Chemistry*, vol. 1, Part 1. Paraffine (38–52), lard (33), butter (33), wax (65), might be used for the solid; alcohol (79), carbon disulphide (48), ether (35), for liquids. For tables of boiling and melting points, see Chute, pp. 368, 369. General discussion, Stewart, pp. 86–158.

REFERENCES.—Avery's *Principles*, 37–42, 368, 372, 374; *Elements*, 496, 502–507, 510–513. Gage's *Introduction*, 110, 111, 113; *Elements*, 128 (p. 160), 130. Hall and Bergen, 169, 170, 185.

Exercise 6. This is a preliminary to specific heat, and may be used as a specific heat determination. The analogy between the relations of specific heat and heat capacity and density and specific gravity generally aids in giving a clear idea of what specific heat really is.

REFERENCES.—Gage's *Elements*, 139, 140.

Exercise 7.—With a glass calorimeter the radiation error is very slight. It might be well to make sure that the students understand the calculation before making the determination. This might be done by giving them imaginary (preferably incorrect) data to work up at some time previous to the exercise. We use cast-iron balls weighing about 130 g., from which we expect from .115 to .120. Glass, lead, ivory, etc., are also occasionally used. Results are best checked by the work of the instructor or by the average results of the class. For tables, see Chute, p. 370. General discussion, Stewart, pp. 285–303.

REFERENCES. Avery's *Principles*, 376, 387–390; *Elements*, 514, 531–536. Gage's *Introduction*, 114; *Elements*, 132, 141–143. Hall and Bergen, 160–163.

Exercise 8.—Results depend largely on care used: 530–545 can be obtained if the student is careful. If desired, the weight of the calorimeter and coil might be marked on the glass once for all, in order to save time. We obtain our best results with about 3500 g. of water. The chief error seems to be inaccurate reading of the thermometer, and insufficient stirring. In absence of litre flasks, the calorimeter could be filled previous to the exercise and the temperature brought down to about 10° by ice or snow, or glass-stoppered bottles containing known volumes might be used. Covering the top of the calorimeter with a piece of card-board, having openings for the steam-pipe and paddle, might tend to check any loss due to evaporation.—Part II. Calculation: The same as for Part I, W being taken as the water put into the calorimeter plus one half the condensed steam. General discussion, Stewart, pp. 304–311.

REFERENCES. Avery's *Principles*, 377–385; *Elements*, 515–529. Gage's *Introduction*, 115–118; *Elements*, 132 (paragraph 2) –138. Hall and Bergen, 171, 172, 186, 188.

Exercises 9, 10, and 11.—Alternatives: Exercise 9. Examples: Different students with different apparatus. Brass, first method, 00001800 and 00001669; second method, 00001809 and 00001785. The ice-water may be dispensed with and the first temperature of the rod taken as the temperature of the room. The use of ice-water, however, impresses on the mind the fact that the length at 0 is that actually used in the determination of the coefficient, and gives a fixed value to start from. The apparatus for the second method might be used for an additional exercise in Mensuration, the student using it as a micrometer-screw to get thickness of sheet-iron, sheet-brass, etc.—Exercise 10. Example (0° to 24°), .0093 theory, 1°, 001049.—Exercise 11. Might calibrate apparatus previous to exercise. The value of 1 cm. on the tube and the volume of vessel A may be marked upon the apparatus. Fair values, 00365, 00367; second method, somewhat less accurate. For tables, see Chute, pp. 368, 369. Brass, .00001875. General discussion, Stewart, pp. 25–84.

REFERENCES. Avery's *Principles*, 366; *Elements*, 492–494. Gage's *Elements*, 114, 115, 124–128. Hall and Bergen, 144–151, 152, 155–159.

Exercise 12.—Example: Part I. In five minutes, black can 6°, bright can 3.5°, rise in temperature. Part II. Black can 18°, bright can 16°, fall in temperature, in 15 min. 250 g. of water in each can. General discussion, see Stewart, pp. 172–252.

REFERENCES. Avery's *Principles*, 398–404; *Elements*, 545–559. Gage's *Introduction*, 313–316; *Elements*, 357–360.

Exercise 13.—Warm water is used to save time. About 25°. Time might be saved by providing the sugar, etc., in little powders containing the required weights. In Part VI a curve might be plotted.

REFERENCES. Avery's *Principles*, 23; *Elements*, 41, 42. Gage's *Introduction*, 6, 7, 21; *Elements*, 22–25, 36–42.

DYNAMICS.

Working out data by geometry; investigation of laws requiring the comparison of a series of measurements of two independent variables; working out formulæ.

Before taking up the first exercise, the ideas of the class about forces, their effects, the conditions necessary to develop them, and their characteristics, might be brought into definite shape. From their work on Magnetism and on Ampère's Law in Electricity, they should be familiar with magnetic force and electromagnetic force. From weighing and general experience they should know something of the force of gravitation, and their general experience should have made them more or less acquainted with other forces. By going over the exercises on Magnetism, and that on Ampère's Law, with the aid of their general information, all the preliminary ideas required might be brought into shape by discussion in the class.

The main points to be brought out are as follows: That a force is anything corresponding to our idea of a push or a pull; that all we can observe when a force is present is motion; that by this motion we recognize the presence of a force; that two bodies (or parts of bodies) are required for the production of a force; that the action is mutual; that contact is not necessary; that forces are usually named from the conditions under which they are developed; that they vary in strength, direction, and point of application. All this can be brought out by questions on the experiments cited above.

Exercise 1.—Maxwell's *Matter and Motion*, Hall's *Elementary Ideas, Definitions, and Laws in Dynamics*.

References.—Avery's *Principles*, 46–57; *Elements*, 38, 64, 70, 72–78. Gage's *Introduction*, 10–14, 59–61, 63, 65 (paragraph 2), 75, 79, 80; *Elements*, 12, 65–69, 76 (pp. 101, 102). Hall and Bergen, 70, 71, 73, 74, 106–111, 114, 115.

Exercise 2.—The coefficient might be calculated. The average results of a number of trials made on different parts of the board in each case will probably give the best results for purposes of comparison.

References. Avery's *Principles*, 142, 143, 409; *Elements*, 212–214, 65–69, 81. Gage's *Introduction*, 62; *Elements*, 95, 97. Hall and Bergen, 72, 81, 112, 121–124.

Exercise 3.—The author performs Experiments 1 and 2 before the class. Best results when balances are held by nails. That pupils may understand the principle upon which Experiment 3 is conducted, special care must be used that the force which holds the point against the combined action of the other two is not considered as the resultant itself. The accuracy of the diagrams obtained will depend greatly on the care used in laying off the lines, reading the balances, etc.

REFERENCES.—Avery's *Principles*, 133, 135–138 ; *Elements*, 79, 80, 82–92, 163, 166, 198–201. Gage's *Introduction*, 64, 65 (paragraph 1); *Elements*, 70, 71. Hall and Bergen, 75, 77–80, 82.

Exercise 4.—Three pupils. Ex.: B × Bd 104, 57 C × Cd 105, 60.

This exercise is also made the basis of instruction on the centre of gravity, considering the centre of gravity as the point of application of the resultant of all the parallel forces which together make up the weight of a body. See Lodge's Mechanics, p. 114. For an exercise on the determination of the centre of gravity by momenta, see Hall and Bergen, p. 121.

REFERENCES. Avery's *Principles*, 65–73, 109–116, 118–131 ; *Elements*, 107–117, 168–197. Gage's *Introduction*, 66–69, 72–74; *Elements*, 72–76. Hall and Bergen, 83, 84, 86–94, 97–104.

Exercise 5.—Probable error, 1 ft. lb. The greater the angle the less the error seems to be.

REFERENCES. Avery's *Principles*, 92–97, 105–108, 133, 134, 406–418 ; *Elements*, 150–155, 164, 165, 202, 203, 561–578. Gage's *Introduction*, 82–85, 89, 94, 95, 119–126 ; *Elements*, 88–91, 100–105, 144–150. Hall and Bergen, 125, 126, 128, 193–198.

Exercise 6.—It may be well to emphasize the fact that work is reckoned by multiplying the force by the distance measured in the line of the force ; in this case Dh. Errors about the same as in Exercise 5. For full discussion of Thermo-dynamics, see Stewart, 77, 312–360.

REFERENCES. Avery's *Principles*, 139–141 ; *Elements*, 205–211.

Exercise 7.—Pupils may need help in getting out the formula. Example where L : L′ = 1 : 2, V : V′ = 1 : 1.34. A curve might be plotted.

REFERENCES. Avery's *Principles*, 59–64, 75–82, 84–90, 98–103 ; *Elements*, 98–106, 118–136, 137–149, 156–162. Gage's *Intro-*

duction, 76–78, 81 ; *Elements*, 18–21, 77–84, 92, 98, 99. Hall and Bergen, 113, 117, 129–133.

Exercise 8.—For preliminary discussion, see Hall and Bergen, p. 143. Three students can work well together. The sums of momenta after collision are usually a little less than those before. The weights of the balls may be determined once for all. For greater accuracy, one half the mass of the suspending wires should be added. The substitute experiment is somewhat less accurate.

REFERENCES. Avery's *Principles*, 54–57, 103 ; *Elements*, 70–72, 93–97, 162. Gage's *Introduction*, 70, 71 ; *Elements*, 85–87, 93, 94. Hall and Bergen, 118, 119.

Exercise 9.—It may be well to caution students against getting their faces so close to the balance as to be injured by the flying up of the hook when the wire breaks.

REFERENCES. Avery's *Principles*, 32 ; *Elements*, 48. Gage's *Introduction*, 20 ; *Elements*, 32. Hall and Bergen, 7, 9, 25.

Exercise 10.—The stretch will increase as the loads are made greater owing to the reduction of cross section. Example : load of 50 g., stretch, 0.8 cm. ; load of 550 g., stretch, 10.47 cm. A curve might be plotted from the data. For discussion, see Stewart and Gee, vol. 1, p. 170.

REFERENCES. Avery's *Principles*, 27 ; *Elements*, 45. Gage's *Introduction*, 24 ; *Elements*, 28. Hall and Bergen, 10, 11, 14–16.

Exercise 11.—Owing to the amount of mercury required, the author alternates this exercise with Exercises 9 and 10, so that but four sets of apparatus are used for a section. Examples, where $V = 1 : 1.10 :: 1.31 : 1.71$; $P = 1 : 1.14 :: 1.30 : 1.72$.

REFERENCES. Avery's *Principles*, 179 ; *Elements*, 45, 282–287. Gage's *Introduction*, 28, 37–39 ; *Elements*, 50.

LIGHT.

Exercise 1.—Example : average of five trials, $1/AL + 1/LS' = 0.05637$ $1/F = .055$.

REFERENCES. Avery's *Principles*, 420–425, 442–459, 472–481, 427 ; *Elements*, 579–584, 588, 611–633, 656–666. Gage's *Introduction*, 260–267, 278–292, 317–321 ; *Elements*, 300–308, 315–333, 362–367. Hall and Bergen, 230–232, 249, 261–263, 265, 266, 272–278, 279–285.

Exercise 3.—Chief error in getting exact area. Fair results: example ratios of distances, 1 : 1.36 : 2.22 : 2.55. Ratios of areas were 1 : 1.4 : 2.0 : 2.5.

REFERENCES. Avery's *Principles*, 426, 428; *Elements*, 585–587, 589. Gage's *Introduction*, 268, 269; *Elements*, 309–314. Hall and Bergen, 234–236.

Exercises 2 and 4.—In Exercise 2 the chief error lies in the fact that different candles seldom give flames of equal intensity. Example of results, candles 1 : 2 : 4 : distances were 1 : 1.78 : 2.3. A determination by the author where care was used to get similar flames, gave: candles 1 : 2 : 3. Distances 1 : 1.8 : 3.1.

REFERENCES. Gage's *Introduction*, 270–272. Hall and Bergen, 239, 240.

SOUND.

Exercise 1.—REFERENCES. Avery's *Principles*, 328-332, 338, 339, 346–352; *Elements*, 428–437, 443–447, 454–471. Gage's *Introduction*, 242–259; *Elements*, 276–299. Hall and Bergen, 212–217, 220, 221, 223–225, 228, 229.

Exercise 2.—A still simpler method would be to measure the time elapsing between seeing the smoke and hearing the report of a gun about half a mile from the observer. Example: At 15.5°, Air, 325 m., Carbon dioxide, 283 m. per second; fork used = 528. Velocity of sound at 0, 333 m. per second. Add .6 m. per degree centigrade.

REFERENCES. Avery's *Principles*, 317–321, 325, 326, 330, 333, 340–345; *Elements*, 415–427, 440, 441, 442, 448–453. Gage's *Introduction*, 216–241; *Elements*, 247–275. Hall and Bergen, 199–207, 211.

INDEX.

☞ **Numbers above 204 refer to teachers' edition.**

Lightning Source UK Ltd.
Milton Keynes UK
UKHW010004231118
332797UK00017B/2163/P